4th
edition

Using IBM® SPSS® Statistics
for Research Methods *and*
Social Science Statistics

For my sister, Becky, and my brother, Ben

4th edition

Using IBM® SPSS® Statistics *for* Research Methods *and* Social Science Statistics

William E. Wagner, III

California State University, Channel Islands

Los Angeles | London | New Delhi
Singapore | Washington DC

Los Angeles | London | New Delhi
Singapore | Washington DC

FOR INFORMATION:

SAGE Publications, Inc.
2455 Teller Road
Thousand Oaks, California 91320
E-mail: order@sagepub.com

SAGE Publications Ltd.
1 Oliver's Yard
55 City Road
London EC1Y 1SP
United Kingdom

SAGE Publications India Pvt. Ltd.
B 1/I 1 Mohan Cooperative Industrial Area
Mathura Road, New Delhi 110 044
India

SAGE Publications Asia-Pacific Pte. Ltd.
3 Church Street
#10-04 Samsung Hub
Singapore 049483

Copyright © 2013 by SAGE Publications, Inc.

Printed in the United States of America

Library of Congress Cataloging-in-Publication Data

Wagner, William E. (William Edward)

Using IBM® SPSS® statistics for research methods and social science statistics / William E. Wagner, III. — 4th ed.

p. cm.
Includes bibliographical references and index.

ISBN 978-1-4522-1770-3 (pbk.)

1. Social sciences—Statistical methods.
2. SPSS for Windows. I. Title.

HA32.W34 2013
300.285′555—dc23   2012003919

This book is printed on acid-free paper.

Acquisitions Editor: Jerry Westby
Editorial Assistant: Erim Sarbuland/Laura Cheung
Production Editor: Eric Garner
Copy Editor: Liann Lech
Typesetter: C&M Digitals (P) Ltd.
Proofreader: Barbara Johnson
Cover Designer: Glenn Vogel
Marketing Manager: Erica DeLuca
Permissions Editor: Karen Ehrmann

Certified Chain of Custody
SUSTAINABLE FORESTRY INITIATIVE
Promoting Sustainable Forestry
www.sfiprogram.org
SFI-01268

SFI label applies to text stock

12 13 14 15 16 10 9 8 7 6 5 4 3 2 1

Brief Contents

Preface

This book was written for those learning introductory statistics or with some basic statistics knowledge who want to use IBM® SPSS® Statistics* software to manage data and/or carry out basic statistical analyses. It can also be a useful tool to gain an understanding of how SPSS Statistics software works before going on to more complicated statistical procedures. This volume is an ideal supplement for a statistics or research methods course. Although it can be used with any research methods or statistics book or materials, this book was tailored to complement *Investigating the Social World*, by Russell Schutt (2009), and *Social Statistics for a Diverse Society*, by Chava Frankfort-Nachmias and Anna Leon-Guerrero (2009). It can also be used as a guide for those working with basic statistics on their own. The book provides information for users about some of the important mechanics of SPSS Statistics operating procedures for simple data management along with accessible introductory instructions to statistical operations.

References

Frankfort-Nachmias, C., & Leon-Guerrero, A. (2009). *Social statistics for a diverse society* (5th ed.). Thousand Oaks, CA: Pine Forge.
Schutt, R. (2009). *Investigating the social world* (6th ed.). Thousand Oaks, CA: Pine Forge.

---

*IBM® SPSS® Statistics was formerly called PASW® Statistics.

Acknowledgments

M any thanks go to Jerry Westby at SAGE for his support and vision. Anna Leon-Guerrero and Russell Schutt were especially helpful, providing insightful reviews and guidance for earlier editions of this book.

1

Overview

This book serves as a guide for those interested in using IBM SPSS Statistics software to assist in statistical data analysis—whether as a companion to a statistics or research methods course, a stand-alone guide for a particular project, or an aid to individual learning. The images and directions used in this book come from IBM SPSS Statistics Version 20.0, first released in the autumn of 2011. If you are using IBM SPSS Statistics version 19, you will notice considerable consistency. For anyone using PASW Statistics 18 or earlier, there will be a great deal of consistency with these instructions and images, although there will be some areas where there are differences due to the upgrades in the SPSS Statistics Version 20 software.

What's the Difference Between SPSS Statistics and PASW Statistics? None

There is essentially no difference. The program formerly known simply as "SPSS" became "SPSS Statistics" with the Version 17.0 release, and then "PASW Statistics" with the Version 18.0 release. After SPSS, Inc. became an IBM company in October 2009, the branding going forward was changed, so that future releases of the software (Version 19.0 and beyond) are known as "IBM SPSS Statistics." When SPSS was originally developed, it stood for Statistical Package in the Social Sciences. The motivation for the PASW (Predictive Analytics Software) branding change was to reflect the considerable reach of the software to more business-oriented realms, although this name is used only for Version 18.

Statistical Software

The SPSS/PASW Statistics software works with several kinds of computer files: data files, output files, and syntax files. Data files are those computer files that contain the information that the user intends to analyze. Output files contain the statistical analysis of these data, often displayed as tables, graphs, and/or charts. Syntax files are computer instructions that tell the SPSS Statistics software what to do. Syntax files are not used with the student version of SPSS Statistics and are dealt with as an advanced application in Chapter 12 of this book. IBM has discontinued the student version of SPSS software for Version 19 and has not released any information about a student version for Version 20 at the time this book was published.

The General Social Survey (GSS) serves the secondary data set used throughout this book to demonstrate typical functions of the statistical software by example. IBM SPSS Statistics is the software program, produced by SPSS, an IBM company as of 2009, in Chicago, Illinois. The GSS is a data set that is read and analyzed by the SPSS Statistics software; it is a data file containing the information to be analyzed. The two things are distinct and can be used in separate contexts without the other, although the GSS data file used for this book, and available on the companion website, is an SPSS data file and cannot be read without opening it in SPSS Statistics or converting it to another file format suitable for use in another program.

About the GSS Data

The National Opinion Research Center (NORC) at the University of Chicago administers the GSS. The GSS was started in 1972 and continues today. The data used for the examples in this book come from the latest available completed version of the GSS collected in 2010. According to NORC, with the exception of the U.S. Census, the GSS is the most frequently analyzed source of information in the social sciences. NORC acknowledges that there are at least 14,000 instances where the GSS has been used for articles in scholarly journals, books, and doctoral dissertations. Furthermore, it is estimated that more than 400,000 students annually use the GSS in their work.

The GSS contains many demographic and attitudinal questions, as well as rotating topics of special interest. A number of core questions have remained unchanged in each survey since 1972. This allows for rich longitudinal research about the attitudes, opinions, and demographics in the United States. Topical questions appear sometimes for just one year; other times, they can appear for a period of years. Therefore, the GSS is versatile as a longitudinal data resource and a relevant cross-sectional resource.

To maximize the amount of information that can be collected in this massive interviewing project, the GSS uses a *split ballot design*, in which NORC asked some questions in only a random subsample of the households and asked other questions in other households. Some questions, including demographic items, were asked of all respondents. When we begin analyzing the GSS data, you will notice that some data items have a substantial number of respondents for whom data are marked as missing. For the most part, this refers to respondents who were not asked that particular question as a result of the split ballot design.

Although many items were asked of only a subsample of respondents, you can still take the responses as representative of the U.S. adult noninstitutionalized population, subject to normal sampling error. For more information about how the GSS data were collected, see Appendix B, Field Work and Interviewer Specifications, and Appendix C, General Coding Instructions, in the *General Social Survey 1972–2010 Cumulative Codebook* (Smith, Marsden, Hout, & Kim, 2011).

SPSS/PASW Electronic Files

IBM Statistics 18 (as well as other versions of SPSS/PASW Statistics) uses different file extensions, or endings, and associated icons to signify types of files. For instance, a file named "file.sav" is a data file called "file." The ".sav" is used to signify that this is a data file. Again, data files contain the information that SPSS/PASW Statistics uses to analyze. A file with the extension ".sps" is an SPSS Statistics syntax file, and a file with the extension ".spv" (or ".spo" for older versions of SPSS software prior to Version 17.0) is an SPSS Statistics output file. Output files contain analyses of data, such as charts, tables, and other statistical and data manipulation information. Syntax files contain coded instructions for SPSS Statistics to perform operations on data and produce output. It is not necessary to create, save, or even deal with syntax files for most basic SPSS Statistics functions; therefore, syntax files will be covered only to the level of description and simple use in Chapter 12.

Opening Existing Data Files

To open an SPSS Statistics data file that you already have or have obtained, select the "File" menu, then choose "Open" and select "Data." (For other file types, see the section on importing data from non-SPSS file formats.) At this point, you will need to navigate the disk drives (or network drives or other sorts of storage devices) to locate the data file that you wish to open. Once you locate the file, either double click on it, or click on it once and then click the "Open" button toward the bottom right side of the "Open Data" dialog box.

SPSS Statistics will then open the data file, and you will be presented with the information in a grid format (somewhat similar to a Microsoft Excel or other spreadsheet environment). You have choices about both the way the information is presented and the information you see. For example, you can choose to see the Data View, presented in the following image. Here, you are viewing the actual data. Note that the variables are listed in columns, with each case recorded as a row. The variable "age" has been selected as a reference point. The data in that column tell the ages of each of the respondents.

	abnomore	abpoor	abrape	absingle	acqntsex	adults	advfront	affrmact	age	aged	agekdbm	astro
1	0	0	0	0	2	1	0	4	31	1	0	
2	0	0	0	0	1	1	0	1	23	3	0	
3	2	2	1	2	0	1	0	4	71	0	21	
4	0	0	0	0	9	1	0	4	82	1	19	
5	2	2	2	2	0	2	2	0	78	1	22	
6	0	0	0	0	0	1	0	2	40	3	24	
7	1	1	1	1	0	1	0	0	46	1	26	
8	2	1	1	2	9	1	0	3	80	0	24	
9	2	2	1	2	0	4	8	0	31	3	17	
10	1	1	1	1	0	3	2	4	99	0	99	
11	0	0	0	0	0	5	0	1	31	1	19	
12	1	2	1	2	0	5	0	4	21	0	0	
13	2	2	2	2	0	2	0	1	58	0	19	
14	2	2	2	2	0	1	0	1	36	0	17	
15	0	0	0	0	0	2	0	3	57	1	21	
16	2	8	1	2	0	2	8	0	28	2	25	
17	8	8	1	2	0	1	0	4	80	0	25	
18	0	0	0	0	0	1	0	3	84	2	19	
19	1	1	1	1	0	2	0	0	51	1	24	
20	0	0	0	0	0	1	0	3	35	2	0	
21	0	0	0	0	2	1	0	4	49	3	0	
22	1	1	1	1	0	3	3	4	56	0	26	
23	2	2	1	2	0	1	0	4	88	0	17	
24	0	0	0	0	0	1	0	1	23	8	14	
25	0	0	0	0	0	2	0	2	35	3	22	
26	1	1	1	1	1	1	0	0	26	3	0	
27	2	2	1	2	1	1	0	4	21	0	0	

Now, click on the Variable View tab, which is located toward the bottom left of the screen. (Note that the Data View tab is currently selected; this is the default when opening a new file with SPSS or PASW Statistics 17 and higher.) Although the information looks somewhat different, you are still looking at the same data file. See the following image:

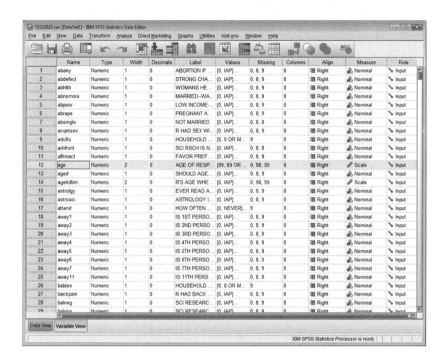

Again, the "age" variable has been selected for reference. In this view, variables are depicted in rows, with each row showing information about a single variable, such as variable label, category label, type, level of measurement, and so forth. You can add to, edit, or delete any of the variable information contained in this view by directly typing into the cells. This view does not show the actual response data; to view that, you would need to select the Data View option.

Importing Data From Statistics File Formats Other Than SPSS or PASW

There is often a need to analyze existing data files that were not created or formatted by SPSS Statistics software. These files might be created by other statistical software packages (e.g., SAS or Stata) or by other types of numeric operational programs (e.g., Microsoft Excel). To open these files, first select the same menu options you would if you were opening an SPSS/ PASW Statistics data file:

File → Open → Data

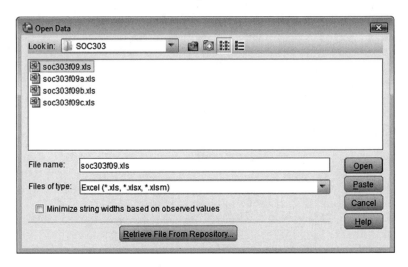

Now, at the "Files of type" prompt at the bottom of the dialog box, click the arrow at the right to expand the choices. Next, select "Excel (*.xls, *.xlsx, *.xlsm)." You will need to navigate your hard drive, other drives, or other locations to find your file. Once you locate the file, select and open it. At this point, you will be presented with a new dialog box:

If the column headings in the Microsoft Excel file contain the variable names, then make sure the box asking to "Read variable names from the first row of data" is selected. If the column headings are not formatted in a way that conforms to the SPSS Statistics variable-naming conventions, then they will be transformed into permitted variable names, and the original names will be recorded as variable labels.

To import only a portion of the Excel file, then enter the range of cells from which you would like to import data.

It is also possible to import files from databases, text files, and other sources. Follow the same instructions as with Excel files, except for the file type you select. Depending on the file type you choose, you will be presented with different dialog boxes or wizards to import the data and process it for use by SPSS Statistics.

In some cases, you may simply have unlabeled data in a particular file, or the variable names or other information may be of little or no use to you. In that case, depending upon the size of the file, you could copy the data from the original numeric program (e.g., Microsoft Excel), and then paste it directly into the SPSS Statistics Data View window. This is particularly useful if you just want to add numeric values from another source and enter or program the other information using SPSS Statistics.

Opening Previously Created Output Files

In order to open a previously created output file, select from the "File" menu as follows:

File → Open → Output

You will be presented with the following dialog box:

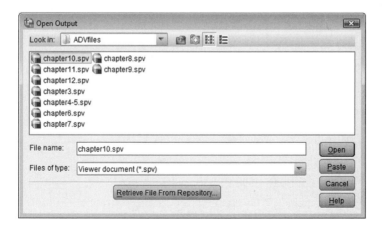

Here, just navigate to find your file, as you would any other type of file. Once you locate it, select and open the file. SPSS Statistics will open the file into an Output Viewer window. There, you can view and edit it. Note that IBM SPSS Statistics Version 20 (as well as Version 19, PASW Statistics 18, and SPSS Statistics 17) uses the *.spv file extension. Earlier versions of SPSS software employ the *.spo file extension. IBM SPSS Statistics Version 20 is capable of opening output files created with older versions of SPSS software, but files created in the new version cannot necessarily be opened in the older versions.

Saving Files

All types of SPSS Statistics files are saved in virtually the same way as files in any modern computer program. Select either

File → Save to save the file as the currently assigned name

or

File → Save As to save the file in a different file, under a new name.

The first option above will automatically save the file without prompting you for a dialog box, unless you are working with a new, yet unnamed, file. In that case, you will get the same type of dialog box as though you had selected the Save As option. If you do choose the second option, you will be given a dialog box prompting you to name the file and to select the location on your computer or network where the file is to be placed.

Creating New SPSS Statistics Data Files

To create a new SPSS Statistics data file, select the File menu; then choose New, and select Data:

File → New → Data

You will then be presented with a blank Data Editor window like the one following.

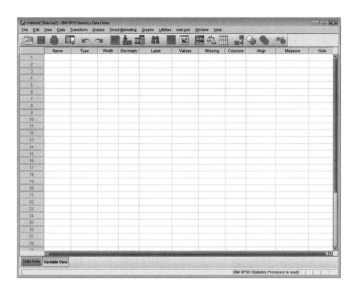

You can immediately start entering information related to the variables you wish to create and/or the actual data codes that you may have. In the Data Editor window that follows, the Variable View tab has been selected and information has been entered for two variables: "age" and "sex." Notice that the labels have been entered: "AGE OF RESPONDENT" and "SEX OF RESPONDENT," respectively. Other information about the variables has been selected and entered, as well.

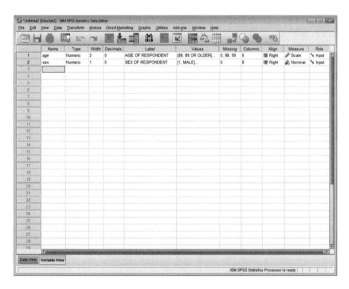

For the "sex" variable, value labels have been entered. This was done by clicking on the "Values" cell for that variable and then selecting the button with three small dots. The following dialog box appeared.

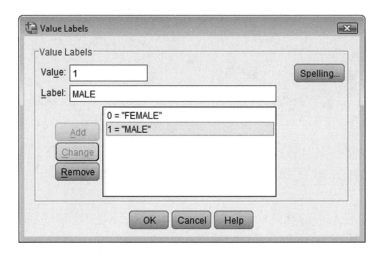

In the Value Labels box, you can enter the label for each of the category codes for the variable. In this case, "0" was entered in the "Value" box, and "FEMALE" was entered into the "Label" box. At that point, to record the information, it is necessary to click the "Add" button. Notice also that "1" was entered into the "Value" box and "MALE" was entered into the "Label" box, and again, the "Add" button was clicked. (This procedure also produces a dummy variable called "male," where the value of 1 is male, and 0 is "not male.") By doing this, we are assigning numbers to the categories of the variable "sex" so that SPSS/PASW Statistics knows how to record whether the respondent is male or female; the software uses numbers to keep track of those attributes. By using numbers, the software can track the categories and use that information to perform statistical operations such as those described throughout this book.

It is also possible to enter the data directly into the Data Editor. To do this, click the Data View tab at the lower left of the Data Editor window. The columns now represent the newly created variables "age" and "sex." You can directly type the values for each case or row. You could also copy (<Ctrl> + C on a PC, or <Apple> + C on a Macintosh) and paste (<Ctrl> + V on a PC, or <Apple> + V on a Macintosh) these values from another software program, such as Microsoft Excel, if you already have them categorized by the same variables.

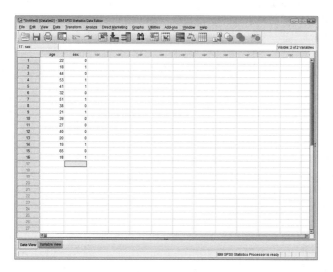

Notice that the data entered (or pasted) above appear as the numeric codes that were assigned for nominal and ordinal variables that have those assignments. There is a way to have the actual label shown in this window. Click the following from the SPSS menus, and note the differences in the screen image below:

View → Value Labels

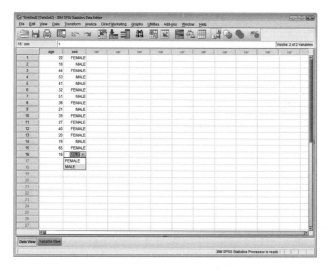

Value labels can also be displayed prior to entering data. When you double-click on the right side of a cell, an arrow will appear. If you click on the arrow, as seen in the screen image above, a pull down menu consisting of all of the available categories for that variable will appear: In this case, the list consists of "FEMALE" and "MALE." If you are entering data directly into SPSS Statistics, using this option can make data entry easier and can help avoid error, such as typos of values that are not within the range of categories for a variable.

Creating and Editing SPSS Statistics Output Files

Output files are created by SPSS Statistics when you instruct the software to perform functions. For example, if you request SPSS Statistics to provide frequencies and central tendency values for three variables from your data set, then an output file will be produced automatically. The information that you have requested will be presented in the Output Viewer window unless an output file is already open in the Output Viewer, in which case the new information will be appended to that file. To edit the output, you'll select and double-click the part you wish to work with, and there are tools to facilitate that task. More information on this topic will be provided in Chapter 4: Organization and Presentation of Information.

Preferences: Getting Started

To change the settings, parameters, and preferences for the SPSS Statistics program, select the Edit menu and choose Options:

Edit → Options

You will be given a dialog box like the one shown here:

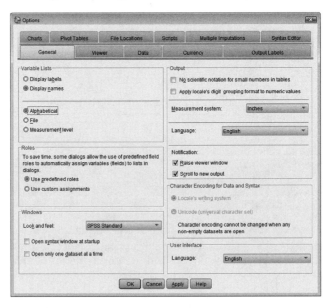

Numerous features can be controlled using this dialog box, and most are intuitive in their operation. As a user becomes more experienced, he or she often uses more of these features. From the start, however, most SPSS Statistics users will want to make sure that variables will be displayed throughout the program in alphabetical order and also by name (rather than label). This can be done by selecting the General tab, and clicking the radio buttons for "Display names" and "Alphabetical."

This is particularly important if you are using or creating a data set that contains a large number of variables, such as the General Social Survey. Although it is clear that alphabetizing the list will facilitate easier access to variables, listing by name is also crucial because variable labels are more detailed and may not necessarily begin with or even use the same letters as the variable name. Changing or verifying these settings up front can save a good deal of time and frustration. If a data set is opened and the preferences have not been set to the desired parameters, the user can still go to the dialog box and make the change while the data set is open. (In some older versions of SPSS, it would be necessary to close the data set, make the change, and reopen the data set.)

If you have started a procedure and don't wish to change the program preferences as described above, you can do it on the fly. Just right-click (or <Control> + Click for Macintosh computers with a standard one-button mouse) one or more of the variables from the list shown in the dialog box with which you are working. You can select "Display Variable Names" and "Sort Alphabetically." Incidentally, you can also select "Variable Information" for any of the variables (not more than one at a time), which will provide details about values, value labels, and so forth.

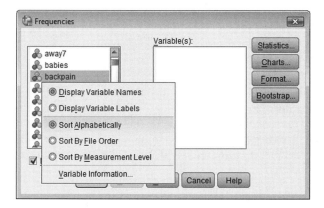

Note that this procedure, pictured above, will make the change only in the current dialog box (including return visits to that same dialog box, "Frequencies" in this case). You will need to use the same procedure to organize variables in other dialog boxes, or make the change at the program preferences window (Edit → Options . . .). Also, note that this procedure does not work in all windows (e.g., Utilities → Variables, see next paragraph).

To get a quick overview of the variables in a given data set, SPSS Statistics has a variable utility window to provide useful information about each of the variables in a way that can be easily navigated (and such that information can be easily pasted to output if desired). Opening this window will demonstrate the importance of organized naming and ordering of variables in a large data file. Choose the Utilities menu, then select Variables:

Utilities → Variables

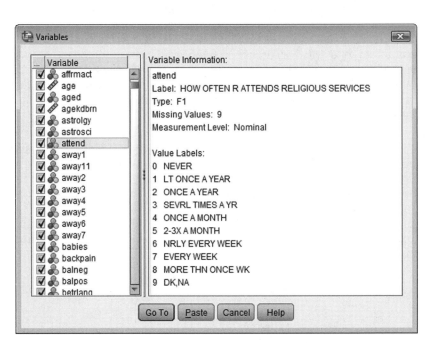

When you select a variable from the alphabetized list of variable names on the left, information about that particular variable will appear on the right side of the box, including the label, the level of measurement, and the value labels. This is a fast way to determine what kind of variables are available in your data set that are suited to different statistical methods of analysis.

Measurement of Variables Using SPSS Statistics

Whether creating a new data file with SPSS Statistics or using an existing data file, it is important to understand how variables have been measured, or "treated," by the creator of the data file. This treatment is a factor of how the data were collected—how much information is contained within the data set about each variable.

First, it is important to be aware that SPSS Statistics can record variables as either *string* variables or *numeric* variables. String variables can consist of letters and/or numbers and cannot be treated numerically; therefore, string variables must be treated at the nominal level of measurement. Numeric variables use numbers to represent response values. These numbers may represent actual numbers, ranked categories, or unranked categories. In other words, numeric variables may be *nominal, ordinal, interval,* or *ratio.*

In social science statistics and research methods courses, variables are typically described using these four categories. Many textbooks, such as *Investigating the Social World* (Schutt, 2009) or *Adventures in Social Research* (Babbie, Halley, Wagner, & Zaino, 2010) elaborate all four of those categories. In some texts, interval and ratio measures are combined, as is the case in *Social Statistics for a Diverse Society* (Frankfort-Nachmias & Leon-Guerrero, 2009).

SPSS Statistics uses the following codes for levels of measurement: nominal, ordinal, and scale. You can select the level of measurement from the pull-down menu for each variable in the "Measure" column of the Variable View window. "Nominal" and "Ordinal" both correspond to the concepts with the same names. The "Scale" denotation corresponds to interval-ratio, interval, and ratio. There are functions within the SPSS Statistics software that will limit your ability to conduct analyses or create graphs based on the recognized level of measurement. Therefore, it is crucial to verify that the indicator in the "Measure" column of the Variable View is correct for all variables you will use in your analyses.

References

Babbie, E., Halley, F., Wagner, W. E., III, & Zaino, J. (2010). *Adventures in social research: Data analysis using IBM® SPSS® Statistics* (7th ed.). Thousand Oaks, CA: Pine Forge.

Frankfort-Nachmias, C., & Leon-Guerrero, A. (2009). *Social statistics for a diverse society* (5th ed.). Thousand Oaks, CA: Pine Forge.

Schutt, R. (2009). *Investigating the social world* (6th ed.). Thousand Oaks, CA: Pine Forge.

Smith, T. W., Marsden, P. V., Hout, M., & Kim, J. (2011). *General Social Survey 1972–2010 cumulative codebook.* Chicago: National Opinion Research Center.

2

Transforming Variables

I n this chapter, tools for restructuring variables will be introduced. IBM SPSS Statistics allows for numerous ways to reconfigure, combine, and compute variables in a data set.

Recoding and Computing Variables

Often, one must reorganize the way data are recorded before performing statistical analyses. This might be due to the level of measurement of a particular variable that a researcher wishes to change, or it could be related to the researcher's intended use of a variable. One may wish to collapse a few categories of a variable into one for appropriateness of analysis. For example, within the marital status variable, one might combine "married" and "separated" categories to form "legally married," and combine "divorced" with "single never married" to form "unmarried." "Recode" is the SPSS Statistics function that allows the researcher to recategorize the variable to suit the needs of the analysis.

There are many times when a researcher needs to produce a new variable from existing information in a data set, but that information is not contained solely within one variable. SPSS Statistics has a "Compute" function that allows a user to both perform mathematical operations on variable data and combine data from multiple distinct variables within the file.

For this recoding example, we will use the General Social Survey (GSS) 2010 data set. We will take a straightforward case of dichotomizing age from a ratio variable, presenting the respondent's actual age at the time of interview, into just two categories with a cut point of 50 years of age. This would allow the researcher to present ages in a small frequency table, whereas before recoding age, there are so many categories (18 to 89 years of age) that presentation in a frequency table is not feasible. In order to recode, or change the categories of a variable, select the "Transform" menu, then choose the "Recode" option, and then select "Into Different Variables":

Transform → Recode into Different Variables . . .

You will then be given a dialog box like the one displayed here:

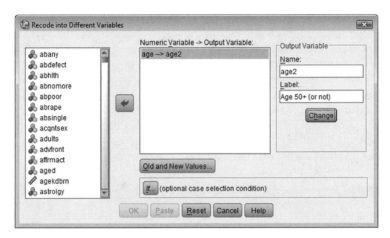

It is useful and proactive against data loss to recode into "different variables," rather than "same variables" if you are reducing information contained within the variable. For instance, if you are recoding a ratio-level variable into an ordinal or a dichotomous variable, then you would want to create a different variable. The reason behind this is that the lost information resulting from the recode would still be retained in the original variable should you want to change the way in which you recode the variable at some point later in time, or should you determine that you need the more detailed ratio-level information for your analysis.

Recoding Variables: Dichotomies and Dummy Variables

In this example, we recode the age variable into a dichotomy: a variable with exactly two categories, not including missing values. A dummy variable is a dichotomy usually coded with a value of "1" to indicate the existence of a particular attribute. All other attributes are coded to "0" as an indication that the particular attribute from "1," and usually the name of the variable itself, is not present. Dummy variables are particularly useful for including nominal-level variables in analyses requiring interval- or ratio-level variables to perform statistical operations, including multiple regression models (see Chapter 7: Correlation and Regression Analysis). It is also possible to use a series of dummy variables to represent several attributes from a given nominal variable.

To dichotomize the age variable, first select "age" from the list of variables in the data set on the left, and then click the arrow to drag it to the "Numeric Variable" pane. Now, create a new name for the variable; in this example, the mundane "age2" has been conjured. You may also select a label at this time, or you can attend to that at a later time through the Variable View tab of the Data Editor screen. For this example, the label "Age 50+ (or not)" was added. Although it has not been done yet in the screen image above, the next step is to click the "Change" button. This will enter the name "age2" into the "Output Variable" location, where there is currently a "?" acting as a placeholder.

Now, it is necessary to give SPSS Statistics instructions for *how* the variable is to be recoded. In this example, we want to change all ages up to and including 49 into a category called "0," and all ages 50 and greater into another category called "1." Click on the "Old and New Values" key in the "Recode into Different Variables" dialog box, and another dialog box will appear on top, like the one that follows:

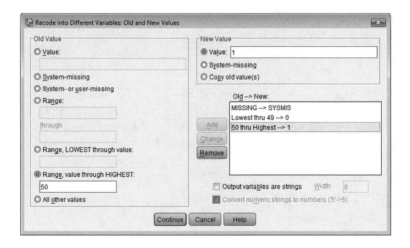

To implement the changes, first, under "Old Value," select the radio button that reads "Range, LOWEST through value." Then enter "49" in the box underneath. Then, under "New Value," select "Value" and enter "0." Now, click the "Add" button. This instructs SPSS Statistics to transform all ages up to and including 49 into the category "0."

Next, under "Old Value," select the radio button associated with "Range, value through HIGHEST." Enter "50" in the box beneath that heading. Then, under "New Value," select "Value" and enter "1." Again, click the "Add" button. This now instructs SPSS Statistics to transform all ages 50 and beyond into the category "1."

Also, under "Old Value," select "System- or user-missing." Under "New Value" select "System-missing." This will ensure that missing values continue to be treated as such, even if they had been recorded as numeric values. Click the "Add" button once again to confirm that this instruction is added to the list.

Now your instructions have been entered, and you can click the "Continue" button; this will close the current dialog box and return you to the original "Recode into Different Variables" dialog box. Once there, you must click the "OK" button for SPSS Statistics to process your request to recode and then create the new variable.

If the "OK" is dimmed and SPSS Statistics will not allow you to click on it, then one of the above steps must not have been completed. The one most often overlooked is the requirement to click on the "Change" button, which adds the new variable name for the output variable.

After you have completed the recode, notice that the new variable is appended to the bottom of the variable list in the Variable View of the Data Editor window. It also happens that the variable is appended to the end (all the way to the right) of the Data View. You can move that variable to another place in the list if you wish by first selecting its entire row and then dragging the row (represented by a red line) up the screen to place it between two of the other variables above. In fact, you can change the "file order" of any of the variables if you wish. This might have utility for you if, for instance, you would like two or more variables to be near each other to compare or verify values in the Data View window.

For the variable information, you may wish to change the width from 8 (the default) to 1, since all possible values are only one digit, including missing case options, such as "IAP" or "Inapplicable" (7), "DK" or "Don't know" (8), and "NA" or "No answer" (9). Similarly, there are no decimals for this variable, so change the decimal column from 2 to 0. Because this variable is dichotomous, select "Nominal" under the "Measure" column.

If you switch to the Data View, you can also see that the variable has been appended at the end of the file (all the way to the right), as shown in the following image.

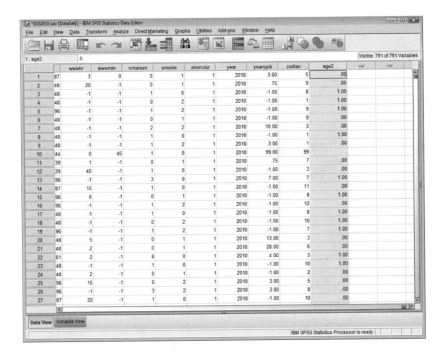

Now, because this is a newly created variable, it is very important to insert value labels. If the variable label and/or value labels are not clear, then it can be easy to forget what the values are, or in which direction the variable was coded, thereby making the data unusable. Click the cell in the "Values" column for the variable to which you want to append value labels. Then click the button with three dots, as shown in the following image.

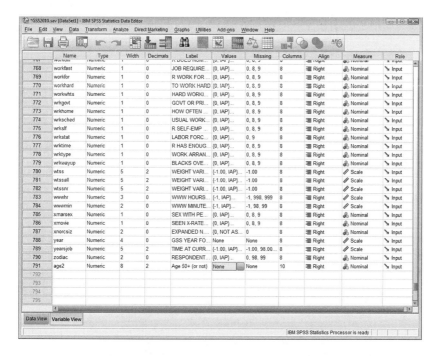

Then, the "Value Labels" dialog box will appear:

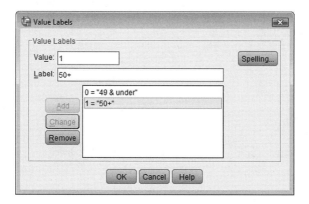

As before with value labels, enter both the value and the label. Then click the "Add" button after each value and label that you enter. When all value/label combinations have been entered, then click "OK" and the value labels will be updated. You can inspect the value labels in the Variable View window by clicking on the button with three dots on it in the "Values" column of the variable "age2."

Recoding Using Two or More Variables to Create a New Variable

Suppose you want to create a new variable of race/ethnicity that includes information from the variable "race," as well as the variable "Hispanic." The variable "race" includes only three categories: "White," "Black," and "Other." We want to select out those who identify as "Hispanic" and create a fourth category labeled as such.

In this example, we are going to recode the race variable into itself using information from the variable "Hispanic." If you would like to preserve the original variable, "race," you can do that by duplicating it first; you might call it "race3" because it consists of three categories.

Let's begin. Select the following menus in SPSS/PASW Statistics:

Transform → Recode into Same Variables . . .

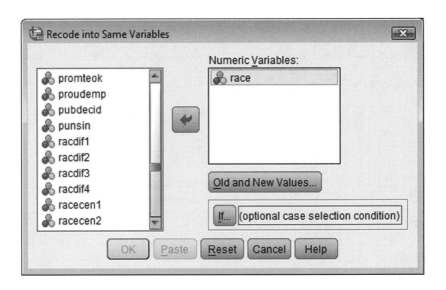

Once the "Recode into Same Variables" dialog box appears, select and move the "race" variable to the "Numeric Variables" slot.

Now, click on "Old and New Values." Here, you will make the following entry: Old values from "1" through "3" will be routed to the new value of "4." See the following image to verify your entry. Then click "Continue."

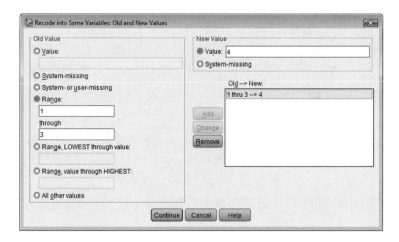

If you think about the logic of what we have just done, we have recoded all respondents (regardless of whether they selected "1," "2," or "3" on the "race" variable) to "Hispanic." But we actually want to recode only the respondents who reported that they were Hispanic. So, let's add that condition to our SPSS Statistics command now. Click the "If" button in the "Recode into Same Variables" dialog box, and you will be presented with the following dialog box:

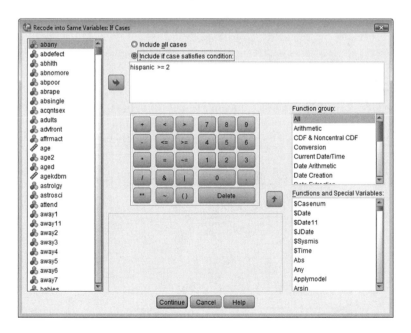

First, select the radio button associated with "Include if case satisfies condition." Now, find the variable "Hispanic" in the list on the left, and click the right-pointing arrow to send it to the slot under "Include if case satisfies condition." Because the variable "Hispanic" is coded "1" for not Hispanic and the values "2"–"50" indicate different types of Hispanic recognition, we want to include those cases where the respondent selected the number "2" or higher. So, click ">=" and then "2." The equation is complete, and you should click "Continue" and then "OK" in the prior dialog box.

Because we have added a new category, it will be necessary to add the additional value label, as illustrated in the following dialog box. Remember that we get this box by going to the Variable View tab of the Data Editor window, finding the cell in the "Value Labels" column for the variable "race," and clicking the button with the three dots.

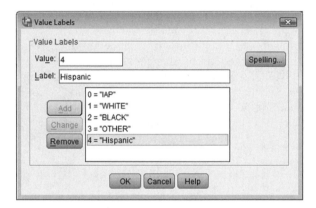

Click "OK" to add the new value label. Now, in order to verify that the recode was done correctly, request a frequency distribution of the new race variable:

Analyze → Descriptive Statistics → Frequencies . . .

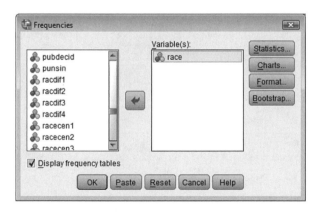

RACE OF RESPONDENT

		Frequency	Percent	Valid Percent	Cumulative Percent
Valid	WHITE	1423	69.6	69.6	69.6
	BLACK	299	14.6	14.6	84.2
	OTHER	87	4.3	4.3	88.5
	Hispanic	235	11.5	11.5	100.0
	Total	2044	100.0	100.0	

You can also ask for frequency distributions of the original race variable, "race3," along with the variable "Hispanic" to verify that the recode was done properly. The distribution above reveals that the recode was carried out as we intended.

Computing Variables

There are numerous reasons why a user of SPSS Statistics would be interested in computing a new variable. For example, one may want to construct an index from individual questions, or one may wish to compute the logarithmic (log) function of a particular variable. In this example, we want to compute the average education level of the respondents' parents. So, we will add the value for mother's education level to that for father's education level and then divide by two. (In more sophisticated approaches, we might divide by the number of parental responses.)

To perform the computation, select these menus:

Transform → Compute . . .

The "Compute Variable" dialog box will be presented:

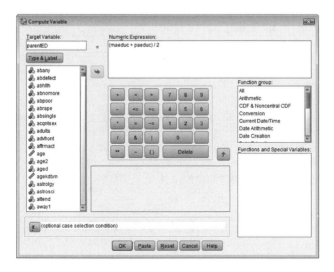

Type the name you wish to assign to the new variable in the "Target Variable" box; in this example, "parentED" was chosen. Next, prepare the computation equation in the "Numeric Expression" box. In this case, it is necessary to select the parentheses button, ( ), first. Then insert the "maeduc" variable by clicking the arrow to the right of the list of variables. Then insert an addition sign by clicking on the appropriate button beneath the "Numeric Expression" box. Now, add the "paeduc" variable into the parentheses as well. Next, insert a divisor bar after the parentheses by clicking the button for such, and then click the number "2." This has the effect of adding together the total years of education of both parents and then dividing by two, yielding the average.

For users who are more comfortable with this process, note that you can type the expression directly into the "Numeric Expression" box that you wish SPSS Statistics to calculate. In this case, you could have typed the expression "(maeduc + paeduc) / 2" using your computer keyboard. If you use this method, be very careful to get the spelling of the variable names exactly correct, or SPSS Statistics will not be able to execute your command.

The "Type & Label" button opens a small dialog box that offers an immediate opportunity to enter a variable label and also to define the type of variable (e.g., numeric).

Note the functions that are available: Each selection in the "Function group" pane will bring up a separate list of functions in the "Functions and Special Variables" pane in the lower right corner of the "Compute Variable" box. There are statistical, trigonometric, date, time, string, and other functions that can be used to compute just about anything. Also, if you wish to set up a conditional computation—such that a computation is made in only one case, or there are to be different computations for different cases based on some predetermined condition, then select the "If" button in the lower left corner and enter that/those condition(s). The same functions and keypads are provided to instruct SPSS Statistics how to determine the criteria for the conditional computation.

Using the Count Function

SPSS Statistics allows users the option to add up particular values across different variables. Suppose a researcher wanted to count up the number of instances where a respondent gave a "yes" answer to particular questions. For this example, consider the General Social Survey (2010) series of questions on opinions relating to abortion. Use the menus below to carry out this example:

Transform → Count Occurrences of Values within Cases . . .

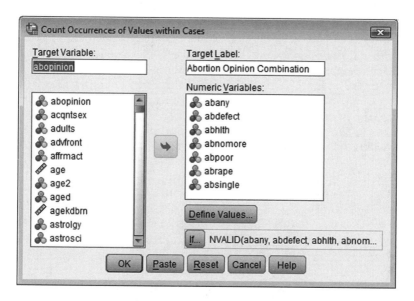

In the dialog box that appears, move the appropriate variables from the variable list on the left into the "Numeric Variables" box. It will also be necessary to enter a name for the new variable to be created in the "Target Variable" slot. The target label can be conveniently entered in the appropriate slot as well. Next, click the "Define Values" button. The following dialog box will be provided:

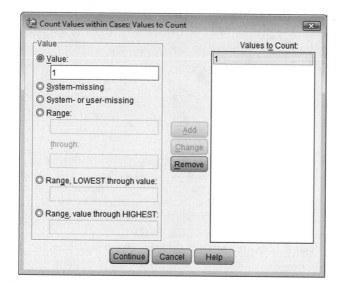

Here, you will want to select the values to be counted. For the opinion questions that have been selected in this example, a code of "1" indicates an affirmative response (and a "2" indicates a negative response). Therefore, you want to count the number of responses that have a value of "1." Click the radio button next to "Value" at the upper left of the dialog box, and then enter a "1" into the associated slot beneath. Now, click the "Add" button in the middle of the dialog box. The "1" should appear in the "Values to Count" area. Now, click the "Continue" button in this box, and you will be taken back to the "Count Occurrences of Values within Cases" dialog box.

We need to make sure that someone who didn't answer one or more of these six questions—or who didn't answer any of them—is not counted as someone who "disapproves." That would corrupt our data, and the results from this analysis would not be valid. We can prevent this by using a conditional filter: Click the "If" button, and the following dialog box will appear.

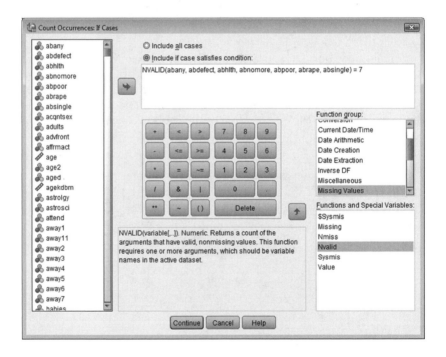

First, select the radio button at the top of the box that corresponds to "Include if case satisfies condition." Next we will prepare the condition. Select "Missing Values" from the "Function group" on the right side of the box. Now, choose "Nvalid" from the list of "Functions and Special Variables." Click the arrow to send it to the box under "Include if case satisfies condition." Then, in the column on the left, find all seven variables having to do with opinions about abortion (abany, abdefect, abhlth, abnomore, abpoor, abrape, and absingle) and drag them between the parentheses next to "NVALID," each separated by a comma. This will return a count for each variable that has a valid response. Therefore, if the value returned is a "7," that means that all seven variables contained valid responses for that case. We want all seven items to have valid responses; otherwise, our new variable might include truncated values (e.g., if someone answered only three questions and approved on all three, the new variable would make it look like he or she approved of only three out of seven, or less than half, when the respondent actually approved in all cases for which he or she replied). So, set this expression equal to seven, as displayed in the illustration above.

Now click "Continue" and then the "OK" button in the prior dialog box, to which you will be returned after this one closes. Other than a record of the SPSS commands executed, no output will be generated. A new variable, however, will be created. See the following screen image for the data contained within the new variable that has been created.

	xmarsex	xmovie	xnorcsiz	year	yearsjob	zodiac	age2	parentED	abopinion	var	var
1	0	1	1	2010	3.00	5	0	5.50			
2	0	1	1	2010	.75	9	0	12.00	.		
3	1	0	1	2010	-1.00	8	1	.	.		
4	0	2	1	2010	-1.00	1	1	1.00	.		
5	1	2	1	2010	-1.00	9	1	.	.		
6	0	1	1	2010	-1.00	9	0	8.00	.		
7	2	2	1	2010	10.00	3	0	14.00	7.00		
8	1	0	1	2010	-1.00	1	1	10.50	4.00		
9	1	2	1	2010	3.00	1	0	.	3.00		
10	1	0	1	2010	99.00	99	.	12.00	7.00		
11	0	1	1	2010	.75	7	0	8.00	.		
12	1	0	1	2010	-1.00	3	0	.	4.00		
13	3	0	1	2010	7.00	7	1	.	.00		
14	1	0	1	2010	-1.00	11	0	12.00	.00		
15	0	1	1	2010	-1.00	9	1	.	.		
16	1	2	1	2010	-1.00	12	0	.	.		
17	1	0	1	2010	-1.00	8	1	10.00	.		
18	0	2	1	2010	-1.00	10	1	.00	.		
19	1	2	1	2010	-1.00	7	1	13.00	7.00		
20	0	1	1	2010	13.00	3	0	.	.		
21	0	1	1	2010	28.00	6	0	12.00	.		
22	8	0	1	2010	4.00	3	1	13.00	7.00		
23	1	0	1	2010	-1.00	10	1	10.00	2.00		
24	0	1	1	2010	-1.00	2	0	.	.		
25	0	2	1	2010	3.00	5	0	.	.		
26	3	2	1	2010	3.00	8	0	8.00	7.00		
27	1	0	1	2010	-1.00	10	0	17.50	2.00		

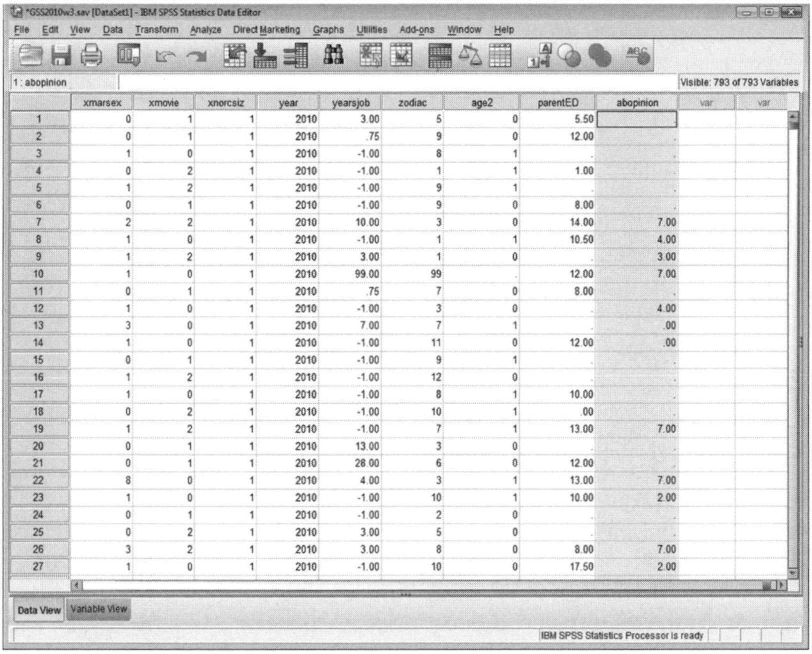

Note that the values, except for those with 0 as the actual value, are all appended with ".00." You will want to go back to the Variable View tab of the Data Editor screen to clean up the details, such as width, decimals, and measure.

Computing an Index Using the Mean

It is possible to construct an index using the "Compute" command in SPSS Statistics. The most direct way of doing this is to use the "Mean" function. To use this method, click these menus:

Transform → Compute Variable . . .

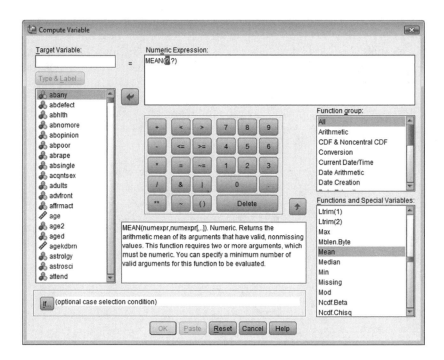

When presented with this dialog box, first note that if any variables, functions, and so on remain from prior use of the Compute Index tool, be sure to click the "Reset" button. Now, in the "Compute Variable" dialog box, select "Mean" from the "Functions and Special Variables" list in the lower right corner. If you don't see "Mean," make sure the "Function group" list above is set to "All." After you select "Mean," click the up-arrow, which will send "Mean" to the pane under "Numeric Expression." It will appear as it does in the preceding image. You must then insert all of the variables of interest within the parentheses after "MEAN" (replacing the question marks), each separated by a comma. See the following dialog box for how this is done in the current example:

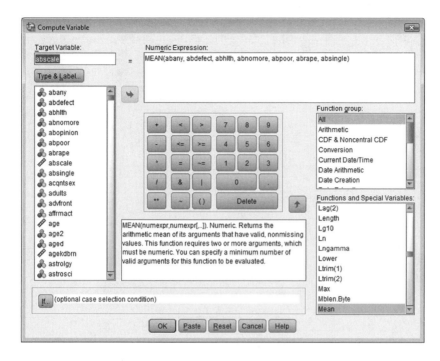

After selecting the variables from the bank on the left and dragging them into the parentheses, enter the name of the target variable (new variable to be created, in this case, "abscale") in the pane under "Target Variable." Then click the "Type & Label" button. The following short dialog box will be provided:

Click the radio button next to "Label" and enter the appropriate variable label in the adjacent slot. Now, click "Continue" and then "OK." A new variable, "abscale," will be created and appear in the SPSS Statistics Data Editor window as shown below.

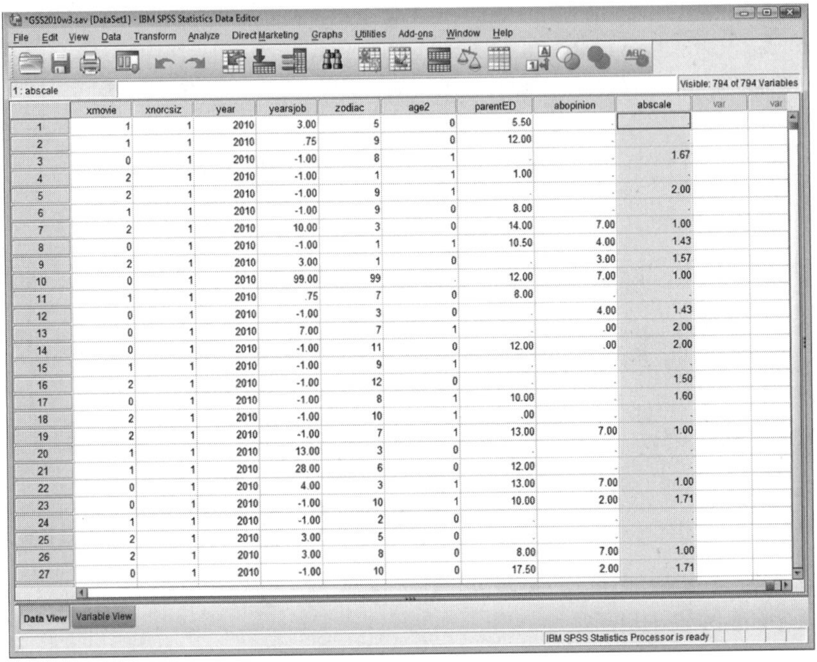

	xmovie	xnorcsiz	year	yearsjob	zodiac	age2	parentED	abopinion	abscale	var	var
1	1	1	2010	3.00	5	0	5.50	.			
2	1	1	2010	.75	9	0	12.00	.	.		
3	0	1	2010	-1.00	8	1	.	.	1.67		
4	2	1	2010	-1.00	1	1	1.00	.	.		
5	2	1	2010	-1.00	9	1	.	.	2.00		
6	1	1	2010	-1.00	9	0	8.00	.	.		
7	2	1	2010	10.00	3	0	14.00	7.00	1.00		
8	0	1	2010	-1.00	1	1	10.50	4.00	1.43		
9	2	1	2010	3.00	1	0	.	3.00	1.57		
10	0	1	2010	99.00	99	.	12.00	7.00	1.00		
11	1	1	2010	.75	7	0	8.00	.	.		
12	0	1	2010	-1.00	3	0	.	4.00	1.43		
13	0	1	2010	7.00	7	1	.	.00	2.00		
14	0	1	2010	-1.00	11	0	12.00	.00	2.00		
15	1	1	2010	-1.00	9	1	.	.	.		
16	2	1	2010	-1.00	12	0	.	.	1.50		
17	0	1	2010	-1.00	8	1	10.00	.	1.60		
18	2	1	2010	-1.00	10	1	.00	.	.		
19	2	1	2010	-1.00	7	1	13.00	7.00	1.00		
20	1	1	2010	13.00	3	0	.	.	.		
21	1	1	2010	28.00	6	0	12.00	.	.		
22	0	1	2010	4.00	3	1	13.00	7.00	1.00		
23	0	1	2010	-1.00	10	1	10.00	2.00	1.71		
24	1	1	2010	-1.00	2	0	.	.	.		
25	2	1	2010	3.00	5	0	.	.	.		
26	2	1	2010	3.00	8	0	8.00	7.00	1.00		
27	0	1	2010	-1.00	10	0	17.50	2.00	1.71		

In this case, for non-missing data (Valid Cases), notice that the value computed for the mean for each subject falls between (and includes) 1 and 2. This is because 1 represents "yes" and 2 represents "no." For statistical purposes, it is sometimes beneficial to have a result between 0 and 1 instead. To arrange for this, you could add a command to subtract the number 1 from the completed mean function in the original "Compute Variable" box, or you can go back now and make the change, as follows. Again, click these menus:

Transform → Compute Variable . . .

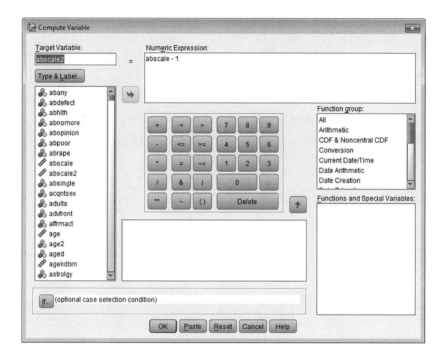

If the dialog box presented is populated with data, click the "Reset" button at the bottom of the box. Now, enter a name for the new variable. (Although it is not recommended, you could overwrite the original variable name.) Click "Type & Label . . ."

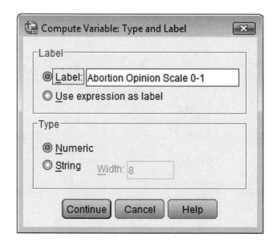

Drag the original variable from the bank on the left to the "Numeric Expression" area. Then, on the calculator-style keypad, click "−" and then "1." This will subtract one from each case and move the data means into the desired range (between 0 and 1), as they are in the following screen image.

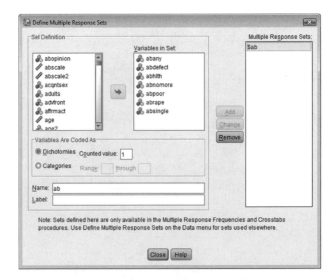

Multiple Response

To produce multiple response values (e.g., frequency values combined across multiple variables), choose the following menus:

Analyze → Multiple Response → Define Variable Sets . . .

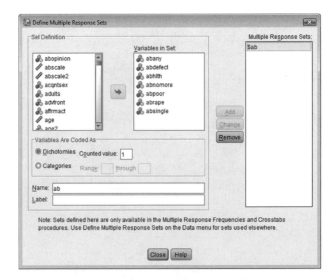

From the variable bank on the left, move all of the desired variables into the "Variables in Set" box. In this example, the variables chosen are dichotomies, and we are counting the "yes" value, 1. It is also possible to select a category and range. Now, name the set, and type the name in the "Name" slot. You also have the opportunity to place a label at this time as well. Finally, click the "Add" button on the right side of the dialog box. You can now click "Close."

In order to produce a frequency table for the response set that has just been identified, click the following menus:

Analyze → Multiple Response → Frequencies . . .

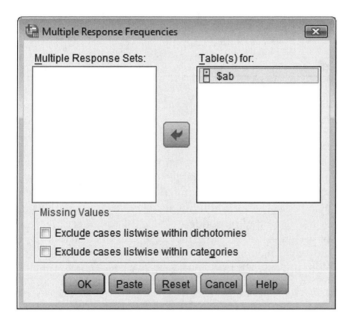

Select the response set from the list on the left. (In this case, it was the only item in the list.) Move the set to the "Table(s) for" box. Then, click "OK." SPSS Statistics will produce output like the following:

Case Summary

	Cases					
	Valid		Missing		Total	
	N	Percent	N	Percent	N	Percent
$ab[a]	1134	55.5%	910	44.5%	2044	100.0%

a. Dichotomy group tabulated at value 1.

$ab Frequencies

		Responses		Percent of Cases
		N	Percent	
$ab[a]	ABORTION IF WOMAN WANTS FOR ANY REASON	537	10.4%	47.4%
	STRONG CHANCE OF SERIOUS DEFECT	921	17.8%	81.2%
	WOMANS HEALTH SERIOUSLY ENDANGERED	1066	20.6%	94.0%
	MARRIED--WANTS NO MORE CHILDREN	587	11.3%	51.8%
	LOW INCOME--CANT AFFORD MORE CHILDREN	560	10.8%	49.4%
	PREGNANT AS RESULT OF RAPE	987	19.0%	87.0%
	NOT MARRIED	524	10.1%	46.2%
Total		5182	100.0%	457.0%

a. Dichotomy group tabulated at value 1.

Note that the multiple response command allows easy production of a table combining similar style variables counting a particular category or range. Above, it is easy to see the similarities and differences in percentages of those who support abortion in the listed circumstances. This frequency table yields the percent of respondents (here called "Percent of Cases") who approve of abortion in the particular case described. Note, for instance, that 94% approve when a woman's health is seriously endangered, whereas only 46.2% approve when the reason is that a woman is not married.

3

Selecting and Sampling Cases

S PSS Statistics software allows a user to draw a sample or subsample from a data set. This selection can be performed to take a random or a targeted sample.

Targeted Selection

For a particular analysis, researchers may not be interested in including all of the cases from a particular data file. There are numerous reasons for this. For instance, if you are interested in studying only characteristics of persons over 21 years of age, then you will need to eliminate any cases in the data set of individuals who are 21 years of age or under.

This condition comes up most often when using a secondary data set, one created by a third party, such as the GSS (General Social Survey). Because the data set was not originally custom tailored to meet your needs, you will need to select the appropriate cases (as well as possibly recode variables, etc., as explained in Chapter 2: Transforming Variables).

To begin, select the following menus:

Data → Select Cases . . .

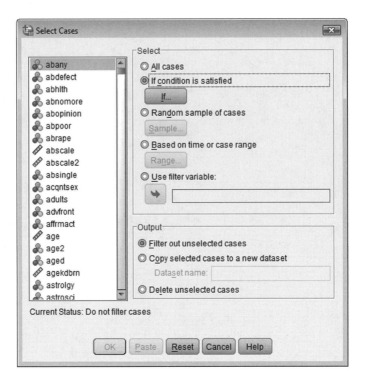

After choosing those menus, you will be given the "Select Cases" dialog box, as pictured. Here, you can choose the types of respondents you would like to analyze and ignore those who do not fit your criteria for inclusion. To do so, choose the radio button next to "If condition is satisfied," and then click on the button labeled "If . . ." underneath it. Once you do that, you will be presented with the "Select Cases: If" dialog box, like the one that follows:

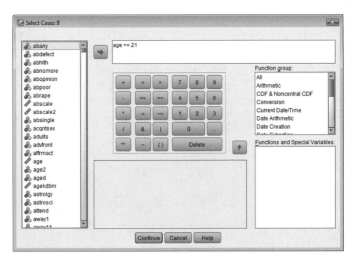

The functions in this box are similar in nature and operation to the functions in the "Compute" dialog box used to compute new variables from existing data, which was described in Chapter 2. Suppose, as described earlier, that you wish to include data only from those respondents who were 21 years of age and older.

First, select the necessary variable(s) from the variable bank on the left. In this case, we need the variable "age." Click the arrow to move it to the right side of the dialog box. Now, click the "> =" (greater than or equals) sign, and then enter the number 21. Alternatively, you could have selected the ">" (greater than) button and entered the number 20. Because age is measured in whole years in the GSS, in this instance, these two methods will both return the same outcome.

Click the "Continue" button in the "Select Cases: If" dialog box, and then click "OK" in the "Select Cases" dialog box. Next, see the Data View window of your data set:

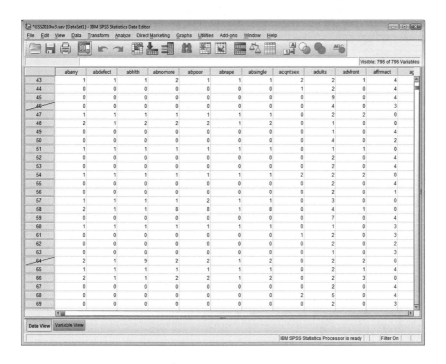

Note that some of the cases have a diagonal line through the SPSS Statistics case/line number at the left side of the window. This is how SPSS Statistics lets you know which cases will be omitted from any and all analyses performed, until the "Select Cases" function is changed or turned off. To turn it off, simply revisit the Data → Select Cases dialog box and either select "All cases" or click the "Reset" button; then click "OK."

In the above example, all cases through #45 were included, so to see which ones were omitted, it was necessary to scroll down (as above). Also, note that whenever a case selection is in effect, SPSS Statistics will alert you with a notification in the lower right-hand corner of the window, indicating "Filter On."

Random Selection

Now, suppose you want to select a random group of cases from a particular data set. This can also be done by calling up the "Select Cases" dialog box:

Data → Select Cases . . .

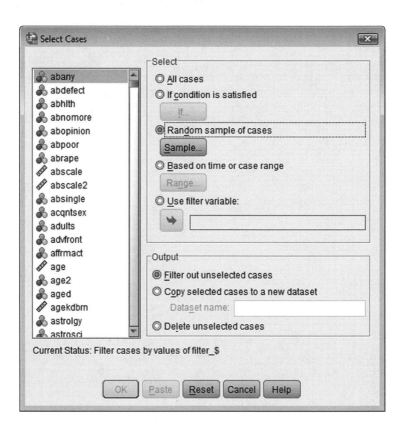

This time, click the "Sample" button after choosing the corresponding radio button "Random sample of cases" in the "Select" section of the dialog box. You will then be given the "Select Cases: Random Sample" dialog box.

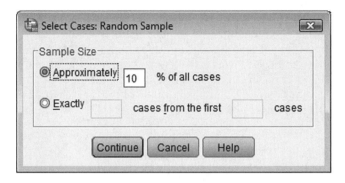

For this example, 10% of the cases will be chosen. SPSS Statistics uses the word *approximately* because not all data sets are necessarily divisible by the percentage that you choose, and hence would not return a whole number of cases in the sample. Note that alternatively, you could select the other radio button and choose an exact number of cases from the first *N* number of cases, as the file is sorted. (Use the Data → Sort Cases menu to change the way the cases in your data set are ordered in SPSS Statistics.)

Click "Continue," then click "OK" when you are taken back to the "Select Cases" dialog box. The image below shows the Variable View tab in the Data Editor window after this action was performed.

Many (approximately 90%) of the cases have a diagonal line through the SPSS Statistics case number at the left. Again, this is how SPSS Statistics lets you know which cases will be omitted from any and all analyses performed, until the "Select Cases" function is changed or turned off. Note that the "Filter On" alert in the lower right-hand corner of the Data Editor window reminds you that only a subsample of the data set will be used for any and all analyses requested.

Selecting Cases for Inclusion in a New Data Set

Up to this point, we have selected cases from a data set while having SPSS Statistics ignore those cases that were not selected for inclusion. It is also possible to have SPSS Statistics create a new data file that contains only the cases that have been selected.

The process of case selection is identical. The difference is in the "Output" section of the dialog box:

Data → Select Cases . . .

Choose the radio button for "Copy selected cases to a new dataset." Then enter a name for your new file; here the name is "subsample." The file will be stored in the default directory as an SPSS Statistics *.sav file. The advantage to choosing this option is that you can work with the subset data file without risk of altering the full original data file. The resulting Data Editor window is shown below.

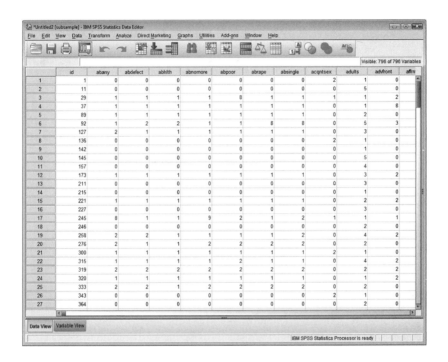

Note that the file has already been assigned the name "subsample" that you provided in the "Select Cases" dialog box; the name appears at the top of the Data Editor window. Also, pay special attention to the case identification variable "id." Because the GSS maintains the "id" variable, you will always be able to track case numbers in the event that you sort cases in a different order. In this instance, you will know which cases were randomly selected from the original GSS data set for this subsample by looking at the "id" variable.

4

Organization and Presentation of Information

In this chapter, basic methods of data description will be exhibited. This information will include frequency distributions, measures of central tendency, and measures of variability. Presentation can be made in the form of a table, chart, or graph.

Measures of Central Tendency and Variability

In order to quickly produce a table with basic descriptive statistics about a variable or variables, select the following menus:

Analyze → Descriptive Statistics → Descriptives . . .

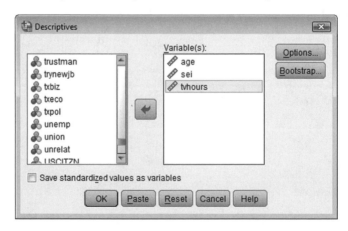

By clicking on the "Options" button at the upper right-hand corner of the "Descriptives" dialog box, you will open another dialog box that will allow you to choose which basic descriptive information will be produced by SPSS Statistics.

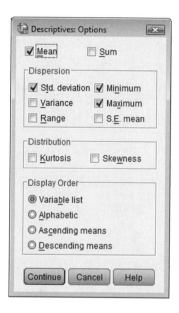

Leave a check mark in the box next to those statistics that you would like to request. You also have the option of choosing the display order; choose one of the four options. After clicking "Continue" in this dialog box and "OK" in the original one, you will be given the following SPSS Statistics output:

Descriptive Statistics

	N	Minimum	Maximum	Mean	Std. Deviation
AGE OF RESPONDENT	2041	18	89	47.97	17.678
RESPONDENT SOCIOECONOMIC INDEX	1875	17.1	97.2	48.992	19.1605
HOURS PER DAY WATCHING TV	1426	0	24	3.03	2.766
Valid N (listwise)	1301				

Note that the sample sizes along with the four measures that were selected for each variable have been presented in separate columns. Variables are listed in the rows of the table output.

Although this method of getting basic descriptive information is very quick and easy, it is possible to get more detailed descriptive information about variables in a data set. Note, for instance, that the previous method will not allow you to obtain the median. You can obtain information about measures of central tendency (including the median) and variability, and you can obtain actual frequency distribution tables using the following method.

Frequency Distributions

Using the same variables as in the previous example, select the following menus:

Analyze → Descriptive Statistics → Frequencies . . .

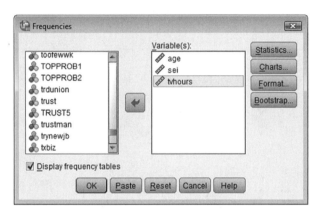

The three selected variables, "age," "sei," and "tvhours," are all scale variables. Therefore, frequency distribution tables for these variables would have too many categories and be too long to be of any real use. You must be cognizant of the level of measurement and categorization of variables before selecting tables. Therefore, make sure that the "Display frequency tables" box is unchecked. After removing the check mark, you will be presented with a pop-up box like this one:

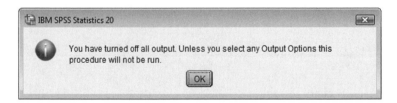

Click "OK" to return to the "Frequencies" dialog box.

To choose which statistical information to request, click the "Statistics" button in the upper right corner of the "Frequencies" dialog box, and you will be presented with the following new dialog box.

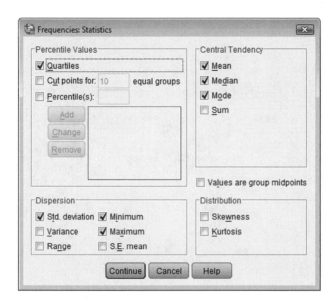

Here, you can choose measures of central tendency (mean, median, and/or mode) and measures of variability. Quartiles are useful for computing the interquartile range (IQR). You can also select any percentile for computation as well, depending on your specific needs. Based on the above dialog boxes, the following output will be provided once you click "Continue" and then "OK" in the original dialog box:

Statistics

		AGE OF RESPONDENT	RESPONDENT SOCIOECONOMIC INDEX	HOURS PER DAY WATCHING TV
N	Valid	2041	1875	1426
	Missing	3	169	618
Mean		47.97	48.992	3.03
Median		47.00	41.200	2.00
Mode		25	63.5	2
Std. Deviation		17.678	19.1605	2.766
Minimum		18	17.1	0
Maximum		89	97.2	24
Percentiles	25	33.00	32.400	1.00
	50	47.00	41.200	2.00
	75	61.00	63.500	4.00

Now, perform the same menu function with a different variable:

Analyze → Descriptives → Frequencies . . .

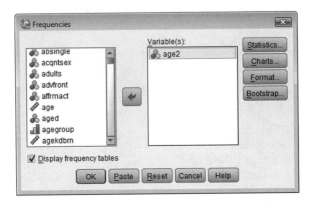

First, remove the variables that were there by using the arrow to send them back to the variable list or by clicking the "Reset" button. Note that if you click the "Reset" button, entries made in the "Statistics" portion of this box will also revert to default, so you would need to click the "Statistics" button and select the appropriate statistics.

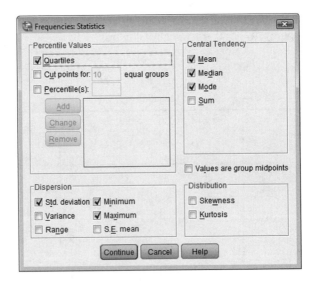

Now, select "age2," the recoded age dichotomy variable (we created this variable in Chapter 2: Transforming Variables), and drag it into the "Variable(s)" area. Next, click the "Charts" button. You will be given a sub-dialog box as follows.

Choose the "Pie charts" radio button and "Frequencies." (Of course, depending on your needs, and the level of measurement of your variable[s], you could select any of these types of charts.) Then click "Continue" and then "OK" in the original dialog box. The information below represents the output that SPSS/PASW Statistics will provide.

Statistics

Age 50+ (or not)

N	Valid	2041
	Missing	3
Mean		.46
Median		.00
Mode		0
Std. Deviation		.498
Variance		.248
Range		1
Minimum		0
Maximum		1
Percentiles	25	.00
	50	.00
	75	1.00

Age 50+ (or not)

		Frequency	Percent	Valid Percent	Cumulative Percent
Valid	49 & under	1106	54.1	54.2	54.2
	50+	935	45.7	45.8	100.0
	Total	2041	99.9	100.0	
Missing	System	3	.1		
Total		2044	100.0		

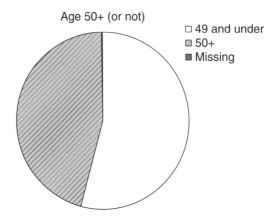

This is only one way to obtain some of the chart and graph options; more detailed options and optimized interfaces to produce charts and graphs are explained in the next chapter. Your chart will likely appear in color and without data labels (percentages in each slice). Instructions for using these features are also addressed in Chapter 5: Charts and Graphs.

Now, suppose that you want to produce a frequency distribution for respondent's age beyond just the dichotomy that was demonstrated in the previous example. It is not feasible to run the frequency command and use its menus using the variable "age" because a virtually useless table listing all ages in the data file would be generated.

In order to present a useful frequency distribution, it makes sense to divide the interval-ratio variable "age" into meaningful or otherwise appropriate categories. Take, for instance, the following example, where "age" is divided into ranges according to decade. Bear in mind that the GSS contains responses only from noninstitutionalized Americans 18 years of age and older. First, use the following menus to recode "age" into "agegroup" (see Chapter 2 for more details on recoding):

Transform → Recode into Different Variables . . .

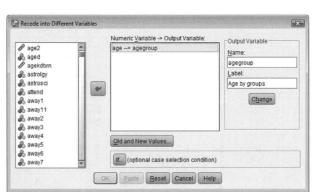

Select the original variable from the variable list on the left, and drag it into the "Numeric Variable → Output Variable" area. On the right side of the dialog box, name the new variable (here, the new name is "agegroup") and provide a label if desired. Next, click the "Change" button to assign the new name that you have entered. Now, click the "Old and New Values" button. The following dialog box will be presented:

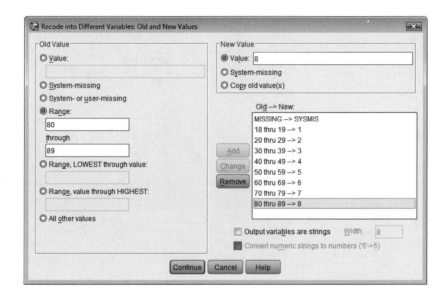

Here, enter the old values and ranges on the left in conjunction with the new value on the right, clicking the "Add" button after each pair of entries. For more details, review the recoding section of Chapter 2: Transforming Variables. After all of the old and new value changes have been added to the list, click the "Continue" button in this window and then the "OK" button in the "Recode into Different Variables" dialog box. The new variable will be displayed in the SPSS Statistics Data Editor window.

Click the Variable View tab and find the newly created variable. Because there are no decimals, change that to "0." Also, there will be no values with more than one digit, so change the width to "1." By recoding the variable, we have transformed it from a scale (ratio) variable into an ordinal variable, so select "ordinal" in the Measure column for the new "agegroup" variable.

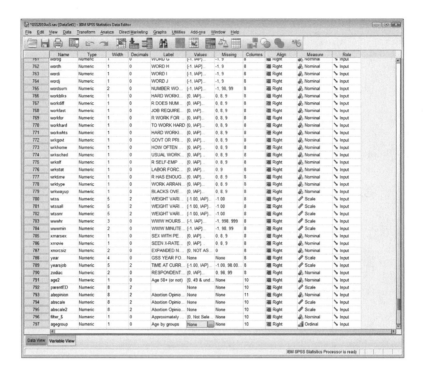

In the "Values" cell, click the button with three dots. You will be given the following dialog box:

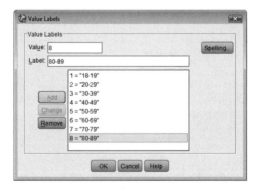

Enter the appropriate labels, as done in the example above. Then click "OK." This will record the value labels onto the variable.

Next, request a frequency distribution for the newly structured variable. To do so, click the following menus:

Analyze → Descriptive Statistics → Frequencies . . .

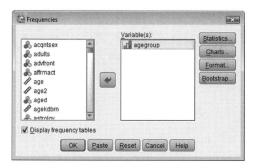

When given the above dialog box, drag the variable of interest from the list on the left into the "Variable(s)" area. For a visual representation of the distribution, select the "Charts . . ." button, and see the dialog box that follows:

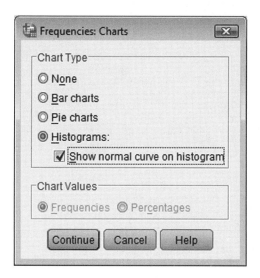

Click the radio button next to "Histograms," and check the box for "Show normal curve on histogram." This will produce a histogram with an overlay of a normal curve for reference. Click the "Continue" button here, and then click "OK" in the prior dialog box. SPSS Statistics will generate the following output, consisting of an easy-to-understand frequency table and a histogram.

Age by groups

		Frequency	Percent	Valid Percent	Cumulative Percent
Valid	18-19	34	1.7	1.7	1.7
	20-29	341	16.7	16.7	18.4
	30-39	363	17.8	17.8	36.2
	40-49	368	18.0	18.0	54.2
	50-59	378	18.5	18.5	72.7
	60-69	302	14.8	14.8	87.5
	70-79	145	7.1	7.1	94.6
	80-89	110	5.4	5.4	100.0
	Total	2041	99.9	100.0	
Missing	System	3	.1		
Total		2044	100.0		

Note that by inspecting the "Valid Percent" column, one can readily get a sense of the distribution, due to the manageable number of categories (rows) in the table. This same information is provided graphically in the histogram that follows:

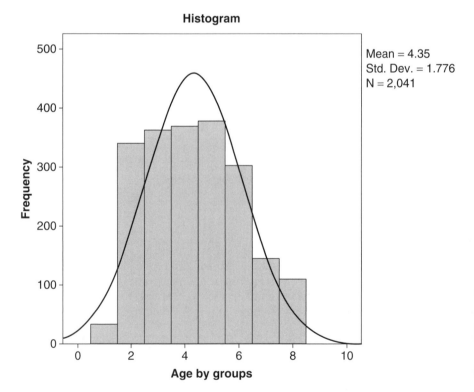

Histogram

Mean = 4.35
Std. Dev. = 1.776
N = 2,041

5

Charts and Graphs

In this chapter, the techniques for producing a number of useful graphics will be explained. There are still other chart and graph options that are not explained, although they are contained within the Graphs menu and can be explored with knowledge of how to use the other IBM SPSS Statistics graph functions along with a background in statistical and research methods.

Most of the charts and graphs discussed in this chapter, as well as those not covered in this book, can be double-clicked in the Output Viewer. After double-clicking, the object opens in a new "Chart Editor" window that allows editing of text, colors, and graphics, as well as the addition of other features and even other variables.

From the SPSS "Statistics Editor" window, the graphs, charts, and tables that are produced can be selected and copied (<Ctrl> + C or <Apple> + C), and then pasted (<Ctrl> + V or <Apple> + V, or "Paste Special") into a word processing program such as Microsoft Word, a publishing program, a web design program, or another program.

Boxplot

A boxplot is a visual representation of the frequency distribution of a variable showing shape, central tendency, and variability of a distribution. It can also be called a box-and-whisker diagram. To produce a boxplot, use the Graphs menu:

Graphs → Chart Builder . . .

If this is the first time you've used the SPSS Statistics "Chart Builder," then you may receive the following alert box.

This alert is an important reminder of at least two things. First, it is important to define the proper level of measurement for any variables for which you will be using this function. Scale (interval and ratio) variables should not be used, for instance, to produce a pie chart. The "Chart Builder" will actually not allow some improper graphing and charting. So, it is doubly important to properly assign levels of measurement in any data set you use, whether it is a secondary data set or whether you collect and enter the data yourself.

Also, make certain that value labels have been assigned properly for nominal and ordinal variables. Without those labels, it may not be possible to make proper use of the information provided by the SPSS Statistics "Chart Builder."

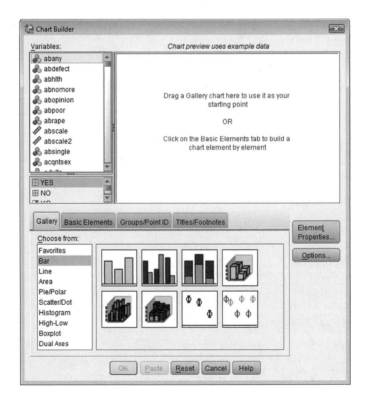

In this example, we will draw a simple boxplot for one variable, "age." From the "Chart Builder" dialog box (see above), select "Boxplot" from the Gallery at the bottom of this window. SPSS Statistics presents three choices for the boxplot, as shown below.

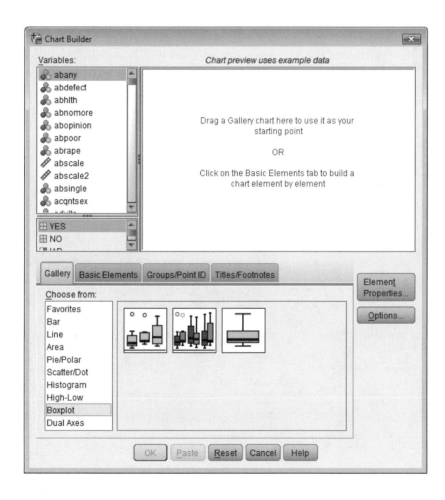

For this example, we will select the third choice, examining just one variable. Drag and drop the icon for your selection into the chart preview area toward the top right side of this window. An area marked "X-Axis?" will appear along the left-hand side of the chart preview pane.

Now, you will need to select a variable from the list on the left to drag into this "X-Axis?" area. SPSS Statistics requires a variable to be dragged to that area in order to compute and draw the boxplot. This example uses the variable "age." Once you do this, the screen will look as follows:

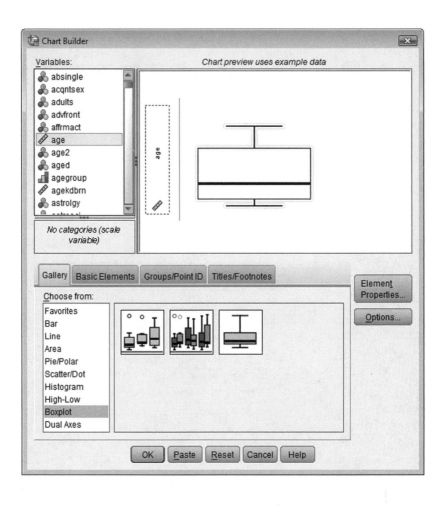

The "Element Properties" window will automatically open. Here, you have options to set the scale range for the graph, as well as the label for the x-axis in this case. (Labels, titles, legends, and so on can all be edited in the output window of SPSS Statistics as well; see Chapter 11: Editing Output.)

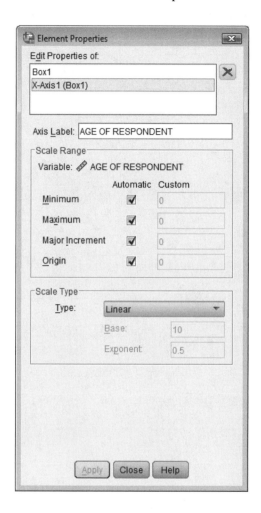

Once you have set your preferences, click "OK" in the "Chart Builder" dialog box, and SPSS Statistics will produce the boxplot in the Output Viewer window. Information about the distribution of the variable is contained in the boxplot. The upper and lower boundaries of the box itself represent Q3 (the 75th percentile) and Q1 (the 25th percentile), respectively. Values correspond to the scale on the left axis of the figure. Therefore, the box itself shows the interquartile range (IQR). The line inside the box is drawn at the 50th percentile (median). The lines extending above and below the box are referred to as whiskers. At the end of the top whisker, the maximum score is marked. The minimum value can be found at the end of the lower whisker.

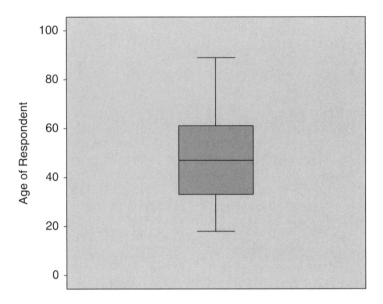

In order to compare the distributions of two subsamples using the boxplot, follow these instructions:

Graphs → Chart Builder . . .

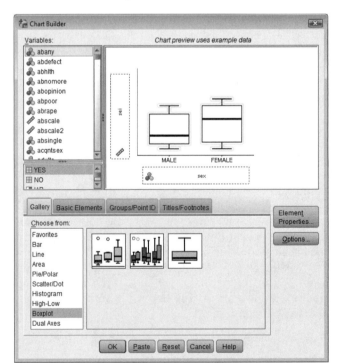

In this case, "sei," or a respondent's socioeconomic index score, is being examined across gender (the variable "sex"). Drag the appropriate variables into the gallery area and click "OK," producing the following output:

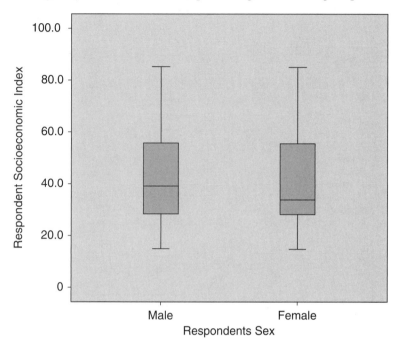

Legacy Options for Graphs (Boxplot Example)

The "Chart Builder" is a newer feature in SPSS Statistics. The option to create charts and graphs using the previous system remains in the program and can be particularly useful for quick creation of certain output, if the researcher knows exactly what type of visual display is desired in advance:

Graphs → Legacy Dialogs → Boxplot . . .

In this example, we will draw a simple boxplot for one variable, "age," as we did before with the "Chart Builder." Click "Simple," and select the button for "Summaries of separate variables." Then click the "Define" button. You will be given another dialog box:

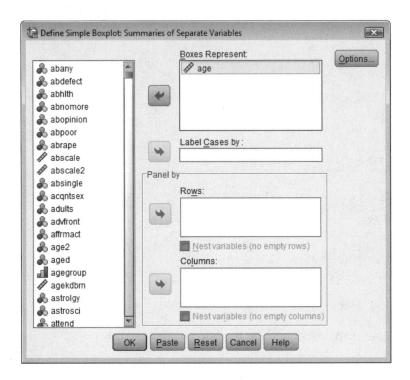

In this dialog box, drag the variable(s) of interest into the "Boxes Represent" location. Click "OK," and SPSS Statistics will provide the boxplot(s) that you have requested. It will be nearly identical to the output provided by the "Chart Builder" option, varying only in ways related to additional formatting options that are selected using the "Chart Builder." A case processing summary is also provided, by default, when creating the boxplot using this method.

Case Processing Summary

	Cases					
	Valid		Missing		Total	
	N	Percent	N	Percent	N	Percent
AGE OF RESPONDENT	2041	99.9%	3	0.1%	2044	100.0%

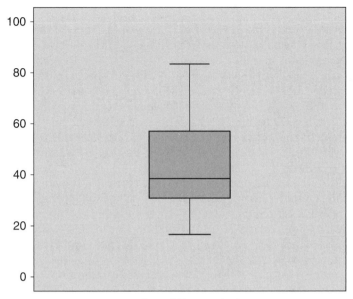

Age of Respondent

Scatterplot

A scatterplot is a useful graph to display the relationship between two scale, or interval-ratio, variables. To create a scatterplot, use these menu directions:

Graphs → Chart Builder . . .

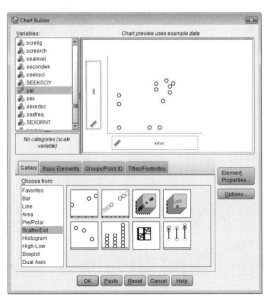

In the "Chart Builder" window, select "Scatter/Dot" from the Gallery. Drag and drop the type of scatterplot you would like to produce into the chart preview area. For this example, choose the first option and drag it to the preview area. Once it is there, you will need to select variables for both the x- and y-axes. For this example, we will plot socioeconomic index by years of education. Drag the "sei" variable from the "Variables" bank on the left into the y-axis location. Then drag the "educ" variable from the "Variables" bank into the x-axis location. Make certain that the levels of measurement for both "sei" and "educ" have been recorded as "scale" (interval-ratio). You can use the "Element Properties" window to change the scale of the graph, edit labels, and so on.

Click "OK" in the "Chart Builder," and SPSS will provide the scatter-plot in the Output Viewer window, as illustrated below. In the graph for SEI by Years of Education, there does appear to be an upward trend. Note also that the data points all fall along vertical lines due to the nature of the variable "educ," which has been categorized into discrete numbers of years (not allowing for 13.5 years, for instance). In the case of a true continuously measured variable, those distinct vertical lines would not show up in the scatterplot.

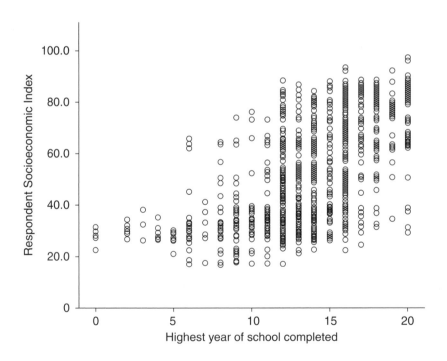

Legacy Scatterplot

To utilize the legacy option, the method still available from prior versions of SPSS software, for creating a scatterplot, use the following menus and proceed using the same information that was entered using the "Chart Builder" option:

Graphs → Legacy Dialogs → Scatter/Dot . . .

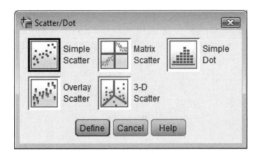

Then, click "Simple Scatter" for our example. (Note that there are several other types of scatterplots available to you, depending upon the nature of the variables you wish to graph.) Click "Define." You will then be presented with the "Simple Scatterplot" dialog box like the one following.

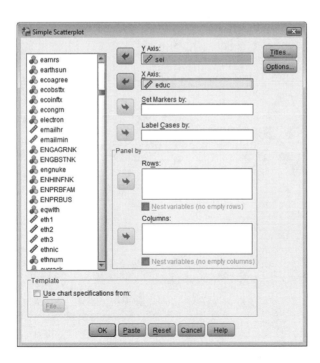

Select variables for the *x*- and *y*-axes. In this example, "sei" (socioeconomic index) has been selected for the *y*-axis (dependent variable), and "edu" (years of education) has been chosen for the *x*-axis (independent variable). Once you click the "OK" button, SPSS Statistics will produce a scatterplot nearly identical to the one provided using the "Chart Builder" option.

Histogram

In order to display a graphic representation of the distribution of a single scale variable, the histogram serves well. This is a bar graph that can be used with variables at the interval and ratio levels. The bars touch. The width of the bars represents the width of the intervals, and the height of the bars represents the frequency of each interval. To create a histogram, follow these menus:

Graphs → Chart Builder . . .

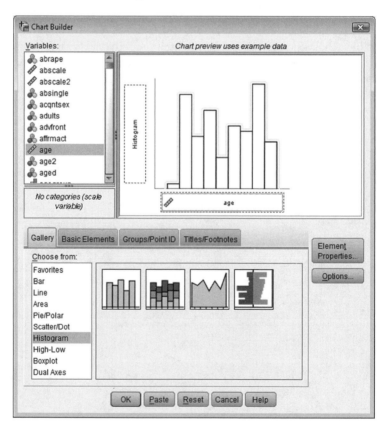

In the "Chart Builder" dialog box, select "Histogram" from the Gallery at the bottom of the window. Then choose the appropriate option; in this example, we will use the first option. Now, drag the variable of interest into the chart preview area. In this case, "age" is the variable selected.

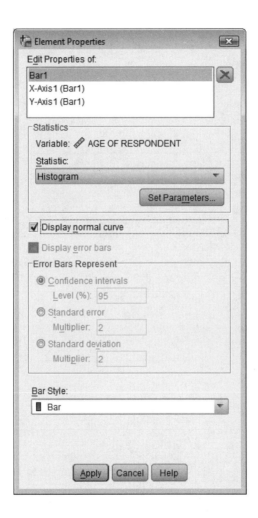

Next, in the "Element Properties" window, making sure that you are editing properties of "Bar1," check the box that is associated with "Display normal curve." Then, click "Apply" in the "Element Properties" window and click "OK" in the "Chart Builder" box.

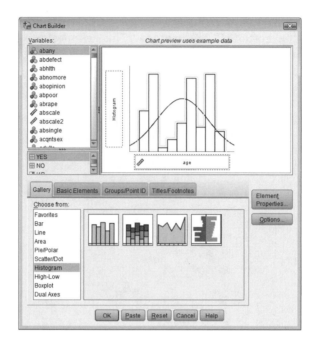

The SPSS Statistics output is shown below. The histogram is displayed, and an overlay of a normal curve has been added, per the request we made in the "Element Properties" window.

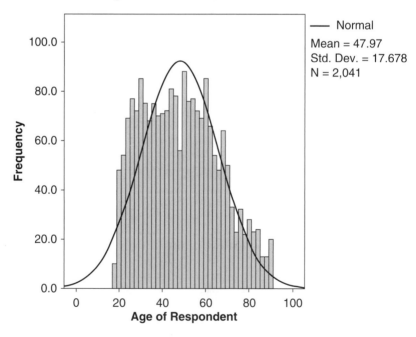

Multivariate Histogram

To produce a stacked histogram, displaying stacked bars of one variable's distribution according to categories of another variable, select the second histogram option in the "Chart Builder" dialog box, the one in which each bar has several different colors. Add the variable of interest, "age," to the x-axis. Then, drag the variable by which you would like the bars divided into the "Stack: set color" box, which will appear in the chart preview area when you drag "age" into it.

Note that you cannot divide bars by a variable that is "scale" according to SPSS Statistics. Even if it is a nominal variable, you cannot move it to the "Stack" area in the chart preview pane unless SPSS Statistics recognizes it as nominal or ordinal. To change a variable from scale to nominal or ordinal (assuming it is indeed not a scale variable), go to the Variable View tab of the Data Editor window. Find the variable you wish to change, click on its cell in the "Measure" column, and then select the new level of measurement from the pull-down choices: nominal, ordinal, or scale.

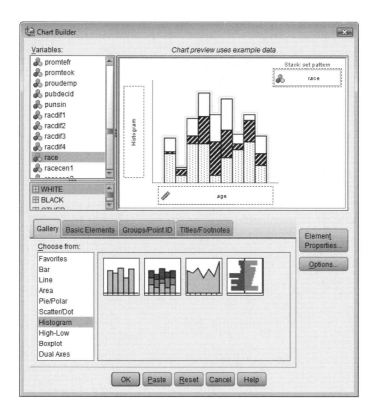

Click "OK," and the following output will be produced in the SPSS Statistics Output Viewer:

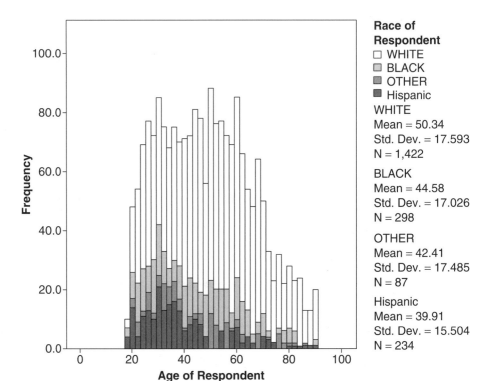

Horizontal Histogram

A two-sided horizontal histogram is another graphic that appears frequently in social science classes, often used to present a "population pyramid." In the "Chart Builder" dialog box, click the fourth choice for the histogram option (the icon furthest to the right), and drag the main axis variable and the variable to define the categories (sides) of the graph into the chart preview area.

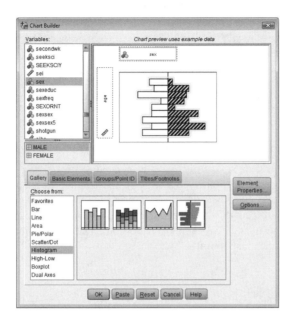

If you use "age" as the distribution variable and wish to divide the distribution by "sex," the following output will be returned after you click "OK":

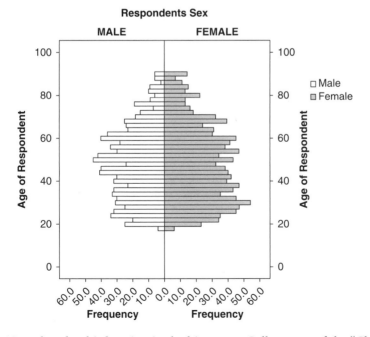

Note that the third option in the histogram Gallery area of the "Chart Builder" is a variation of a frequency polygon that can be created using the same technique used to create the first histogram option.

To produce histograms using an older version of SPSS, or to use that method in the newer SPSS/PASW Statistics versions, select the following menus, and enter the information about your variables and the type of graph you would like to produce:

Graphs → Legacy Dialogs → Histogram . . .

Bar Graph

Usage of bar graphs is common and varied. SPSS Statistics provides many ways of using bar graphs to illustrate information. Much like Microsoft Excel and other spreadsheet and data graphics programs, SPSS Statistics produces bar graphs in a number of different ways. It is possible to create standard bar graphs as well as clustered or stacked bar graphs. To produce a bar graph, use these menus:

Graphs → Chart Builder . . .

Choose "Bar" from the Gallery. You will be given a small dialog box that precedes the main "Bar Chart" dialog boxes. For this example, click "Simple," and then click the "Define" button. Now you will be given the "Define Simple Bar" dialog box and can enter the information to produce the graph.

Be sure that your variable is recognized by SPSS Statistics as having the correct level of measurement. For instance, if you have a nominal variable that is coded in SPSS Statistics as a scale variable, the "Bar" graph option will automatically turn the bar graph into a histogram.

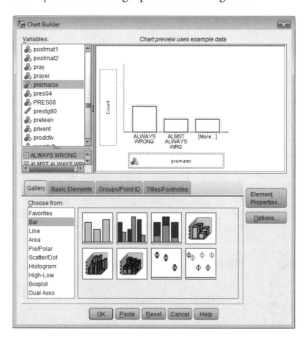

Choose the first bar graph option, and drag it into the chart preview area. Select the variable for graphing from the list on the left, and drag it into the *x*-axis box. Then click "OK." Following is an image of the graph that SPSS Statistics will produce.

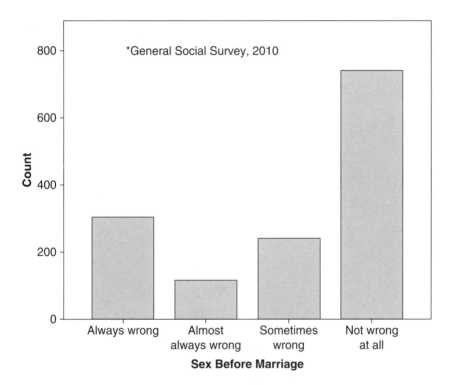

Note that a title and annotation have been added. This can be done by, first, double-clicking the graph in the SPSS Statistics Output Viewer. Then, in the "Chart Editor" dialog box that appears, use the "Option" menu to add a title and/or a text box. The text box can be moved and/or resized. More information about how to add features or edit graphs can be found in Chapter 11: Editing Output.

Multivariate Bar Graph

One might wish to produce clustered bar graphs when comparing information, such as that shown above about beliefs concerning premarital sex, across categories of things like gender, race/ethnicity, or age groups. In the example that follows, we examine beliefs about premarital sex by gender. With a clustered bar graph in this case, it is beneficial to graph the percent of respondents versus the number of respondents, so that the relative bar lengths can be more easily compared within categories of the dependent variable. So, select the following menus:

Graphs → Chart Builder . . .

After selecting "Bar" from the Gallery, select the clustered option, which is the second icon from the left in the top row, as seen in the following image. Then drag it to the chart preview area.

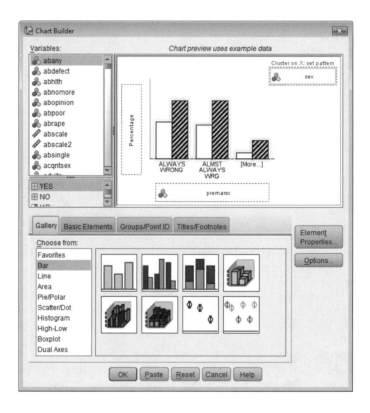

You have the option at this point to enter titles and footnotes by selecting the Titles/Footnotes tab from the row of tabs to the right of the Gallery tab. In this example, you will notice that the output (shown in the figure below) has been edited to include a title and footnote. (You can also add this information later in the SPSS Statistics Output Viewer window by double-clicking the object and working with the editing tools there; for more information, see Chapter 11: Editing Output.)

Now, tell SPSS Statistics with which variables you will be working. Drag the dependent variable, "premarsx," into the x-axis box in the chart preview area. Next, drag the independent variable, "sex," from the variable list into the "Cluster on X" box in the chart preview area.

Next, select "Element Properties" and make sure to change the bar graph option from count to percentage.

Then click on the "Set Parameters" button, and elect the option for percentage that uses all of the bars of the same color (total for that grouping category) as the denominator.

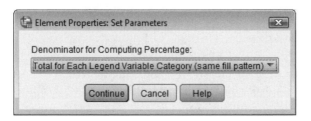

Now, click "Continue" in this box, then "Apply" in the Element Properties box, and then "OK" in the "Chart Builder" window. SPSS Statistics will produce the following output:

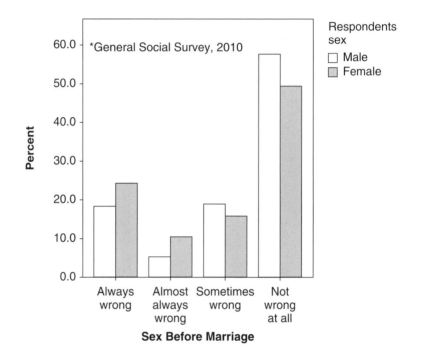

It is apparent from the bar chart prepared by SPSS Statistics that females tend to take a less favorable view of premarital sex than do males. Notice that a larger percentage of women than men in the sample responded "always wrong." Likewise, a smaller percentage of the women responded "not wrong at all."

To produce bar graphs using an older version of SPSS, or to use a legacy method on a newer version of SPSS Statistics, select the following menus and enter the information about your variables and the type of graph you would like to produce:

Graphs → Legacy Dialogs → Bar . . .

Pie Chart

Pie charts are circular graphs with slices that represent the proportion of the total contained within each category. In order to produce a pie chart, select the following menus:

Graphs → Chart Builder . . .

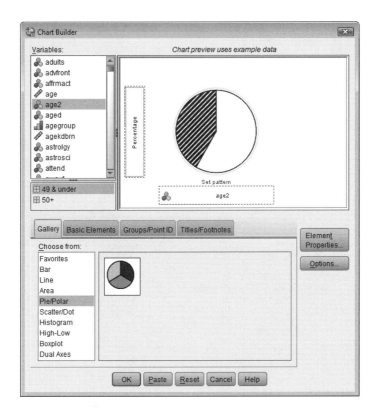

Remember, SPSS Statistics will not allow you to use variables that it recognizes as "scale" to draw a pie chart. If the variable you wish to draw is not scale, you may change the level of measurement recognition for that variable in the Variable View tab of the Data Editor by double-clicking on that variable's cell in the "Measure" column.

For this example, use "age2," the recoded version of respondent's age. "Age2" reduces the data into exactly two categories; this is known as a dichotomy. A pie chart is useful for nominal and many ordinal variables and particularly for dichotomies. It would be of little or no use to create a pie chart for the original "age" variable because there would be too many slices from which to make much sense.

First, click "Pie/Polar" from the Gallery at the bottom of the "Chart Builder" box. Then drag the icon for the only available option to the chart preview area. Now, drag the "age2" variable to the box that appears underneath the sample pie chart in the chart preview area.

Again, note that you have the option here to select the Titles/Footnotes tab, where you will be presented with spaces to easily enter that information. In this example, you will notice that the output (below) has been edited to include a title and footnote. (You can also add this information later in the SPSS Statistics Output Viewer window by double-clicking the object and working with the editing tools there; for more information, see Chapter 11: Editing Output.)

Next, click on "Element Properties" and choose "Polar-interval1."

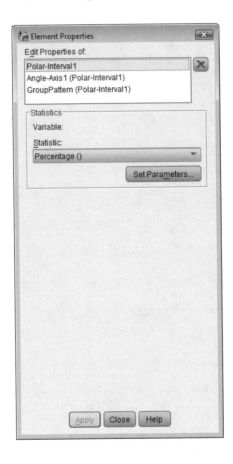

In the "Statistics" box, select "Percentage." There will be only one choice when you click the "Set Parameters" button: Grand Total.

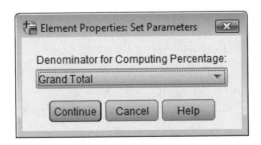

Now, click the "Continue" button, then the "Apply" button in the "Element Properties" window, and then the "OK" button in the "Chart Builder" box. The following graphic will be produced in the SPSS Statistics Output Viewer window:

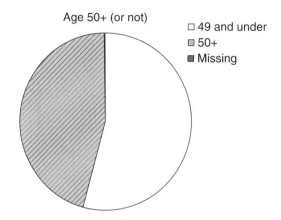

The SPSS Statistics output above provides descriptive (percentage) information in visual form for the variable selected, "age2": age distribution reduced into two categories.

To produce pie charts using older versions of SPSS Software or its legacy method, select the following menus and enter the information about your variables and the type of chart you would like to produce:

Graphs → Legacy Dialogs → Pie . . .

Additional Graphic Capabilities in SPSS Statistics

Using the SPSS Statistics "Chart Builder," you can create additional types of graphics not covered in this chapter, such as line graphs and graphs with dual axes, by selecting those templates from the Gallery in the "Chart Builder" box. The options for patterns, clustering, and so on can be followed from the instructions for the graphs discussed in this chapter.

6

Cross-Tabulation and Measures of Association for Nominal and Ordinal Variables

The most basic type of cross-tabulation (crosstabs) is used to analyze relationships between two variables. This allows a researcher to explore the relationship between variables by examining the intersections of categories of each of the variables involved. The simplest type of cross-tabulation is bivariate analysis, an analysis of two variables. However, the analysis can be expanded beyond that.

Bivariate Analysis

Follow these menus to conduct a sample cross-tabulation of two variables:

Analyze → Descriptive Statistics → Crosstabs . . .

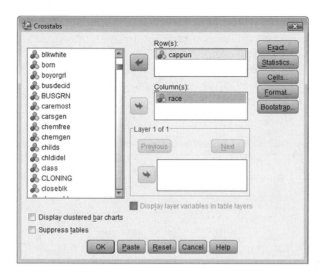

After selecting those menus, you will be presented with a dialog box like the one above. Here, you will have the opportunity to select the row and column variables for the bivariate table. As is customary, it is recommended that you choose the independent variable as the column variable. Above, "cappun" (view on capital punishment/the death penalty) has been selected for the row variable, and "race" (recoded version of respondent's race) has been chosen as the column variable. We use "race" because the number of categories has been limited to four, based on three categories in the original race variable (now "race3") and 50 categories of the variable "Hispanic." It is easier to interpret data from cross-tabulations when the number of categories is kept smaller. Next, click on the "Cells" button to choose options about what information will be given in the output table. The "Crosstabs: Cell Display" dialog box will open.

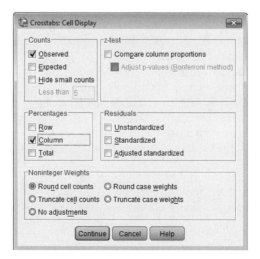

Be sure that the "Column" box is checked under "Percentages." This will make sure that you have information from the appropriate perspective to analyze your variables based on which is the predictor. Click "Continue" in the "Cell Display" dialog box and then "OK" in the original "Crosstabs" dialog box. The table that follows comes from the output produced by following the aforementioned steps:

FAVOR OR OPPOSE DEATH PENALTY FOR MURDER * RACE OF RESPONDENT Crosstabulation

			RACE OF RESPONDENT				Total
			WHITE	BLACK	OTHER	Hispanic	
FAVOR OR OPPOSE DEATH PENALTY FOR MURDER	FAVOR	Count	1002	125	50	120	1297
		% within RACE OF RESPONDENT	74.4%	46.5%	62.5%	53.3%	67.5%
	OPPOSE	Count	345	144	30	105	624
		% within RACE OF RESPONDENT	25.6%	53.5%	37.5%	46.7%	32.5%
Total		Count	1347	269	80	225	1921
		% within RACE OF RESPONDENT	100.0%	100.0%	100.0%	100.0%	100.0%

Based on the information in the table, it is easy to see that there is some sort of relationship between the variables of interest in this case. Note that by looking at the percentages across the columns (categories of the independent variable), you can see that there are differences in opinion about the death penalty by race. According to these GSS data, Whites (74.4%) and Others (62.5%) are more likely to favor the death penalty than Hispanics (53.3%) or Blacks (46.5%).

Adding Another Variable or Dimension to the Analysis

Suppose we want to further explore the bivariate relationship that we briefly examined in the preceding section. By adding another variable, such as respondent's sex, we can further explore how opinions about capital punishment are held in the United States. One way that we can perform this type of analysis is to split our data file by respondent's sex. At that point, any analysis that we do will be performed across the categories of the variable with which we have split the data set.

In order to split the data file by respondent's sex, use these menus:

Data → Split File . . .

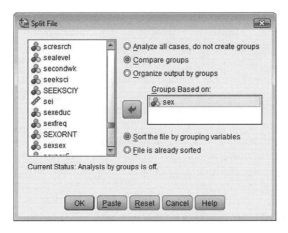

You will be given a "Split File" dialog box like the one above. Here, choose the "Compare groups" radio button. This will brighten the "Groups Based on" box and allow you to drag variables into that box, and these variables will then be used to split the data file. Find "sex" from the variable bank on the left and drag it into the "Groups Based on" box. It is often a good idea to make sure that the file is sorted by grouping variables, although this is not necessary, by clicking the radio button just below the "Groups Based on" box. To continue, click "OK."

SPSS Statistics will now perform the split file function. You will know that the data file has been successfully split by the indicator in the lower right-hand window of the Data Editor screen. It will say "Split by sex."

Now, go back to the "Crosstabs" box, and perform the same operations that were done in the previous section. (The variables and setting should remain the same from before, so unless you've restarted SPSS Statistics in between, it will be just a matter of choosing "OK" once the dialog box appears.)

Analyze → Descriptive Statistics → Crosstabs . . .

FAVOR OR OPPOSE DEATH PENALTY FOR MURDER * RACE OF RESPONDENT Crosstabulation

RESPONDENTS SEX					RACE OF RESPONDENT				
					WHITE	BLACK	OTHER	Hispanic	Total
MALE	FAVOR OR OPPOSE DEATH PENALTY FOR MURDER	FAVOR	Count		484	57	22	51	614
			% within RACE OF RESPONDENT		78.3%	56.4%	64.7%	53.1%	72.3%
		OPPOSE	Count		134	44	12	45	235
			% within RACE OF RESPONDENT		21.7%	43.6%	35.3%	46.9%	27.7%
	Total		Count		618	101	34	96	849
			% within RACE OF RESPONDENT		100.0%	100.0%	100.0%	100.0%	100.0%
FEMALE	FAVOR OR OPPOSE DEATH PENALTY FOR MURDER	FAVOR	Count		518	68	28	69	683
			% within RACE OF RESPONDENT		71.1%	40.5%	60.9%	53.5%	63.7%
		OPPOSE	Count		211	100	18	60	389
			% within RACE OF RESPONDENT		28.9%	59.5%	39.1%	46.5%	36.3%
	Total		Count		729	168	46	129	1072
			% within RACE OF RESPONDENT		100.0%	100.0%	100.0%	100.0%	100.0%

The preceding table will be presented as part of the output that SPSS Statistics returns. Although it is similar to the table given in the previous section, note that it has twice as many cells. It has been split into two tables, one for males and one for females. In this instance, the table shows, among other things, that Black females have the lowest percentage, 40.5%, of all categories of men and women who "favor the death penalty for murder." In all racial categories except "Hispanic," men were more likely to favor the death penalty than women. By adding this new dimension, we were able to obtain some additional insight into public opinion on this matter. See your more comprehensive statistics and/or research methods book(s) for more details.

Measures of Association for Nominal and Ordinal Variables

Proportional reduction in error (PRE) statistics allow us to determine the proportional reduction of error achieved by adding one or more variables to an analysis (even if it is the initial independent variable). "PRE measures are derived by comparing the errors made in predicting the dependent variable while ignoring the independent variable with errors made when making predictions that use information about the independent variable" (Frankfort-Nachmias & Leon-Guerrero, 2009, p. 386). For nominal variables, a PRE statistic that we can use is lambda. For details on how lambda is specifically calculated, see Chapter 12 of *Social Statistics for a Diverse Society* (Frankfort-Nachmias & Leon-Guerrero, 2009) or Chapter 13 of *Adventures in Social Research* (Babbie, Halley, Wagner, & Zaino, 2010).

Lambda (λ)

Lambda is a measure of association for nominal variables. It ranges from 0 to 1. When lambda equals 0, then there is no association; none of the variation in the dependent variable can be explained by the variation in the independent variable. When lambda equals 1, then it is a perfect (deterministic) association; 100% (all) of the variation in the dependent variable can be explained by the variation of the independent variable.

To compute lambda by hand, we would calculate E1 and E2. E1 gives the number of errors that would be made predicting the dependent variable while ignoring the information available in the independent variable. E2 represents the errors of prediction of the dependent variable taking into account the information obtained from the independent variable. Using those errors, lambda is calculated as follows:

$$\lambda = \frac{E1 - E2}{E1}$$

Before computing lambda, turn off the split file feature if you haven't already done so or restarted SPSS Statistics. Open the "Split File" dialog box as follows:

Data → Split File . . .

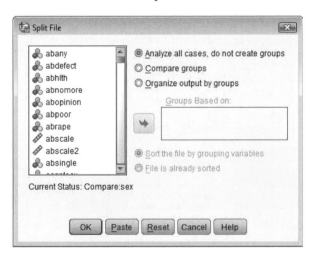

Now, choose "Analyze all cases, do not create groups" and click "OK." SPSS/PASW Statistics will now not split the file, so that all cases in the data set will be analyzed.

To compute lambda for the relationship between race and view on capital punishment, begin again by selecting the cross-tabulation menu:

Analyze → Descriptive Statistics → Crosstabs . . .

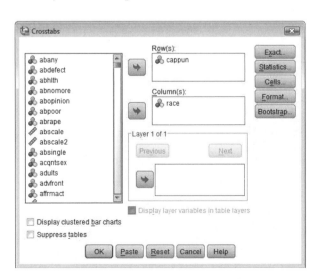

Now, when presented the "Crosstabs" dialog box, and after entering the variables of interest ("race" and "cappun" in this case), select the "Statistics" button. You will be given the following dialog box:

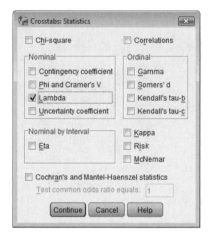

Under the "Nominal" heading, select "Lambda." This will instruct SPSS Statistics to add lambda to the things it will present in the output. Now, click "Continue" in the "Statistics" dialog box, and then "OK" in the prior dialog box. Below is an image from the output SPSS will produce:

Directional Measures

			Value	Asymp. Std. Error[a]	Approx. T[b]	Approx. Sig.
Nominal by Nominal	Lambda	Symmetric	.016	.014	1.159	.247
		FAVOR OR OPPOSE DEATH PENALTY FOR MURDER Dependent	.030	.026	1.159	.247
		RACE OF RESPONDENT ...	.000	.000	.[c]	.[c]
	Goodman and Kruskal tau	FAVOR OR OPPOSE DEATH PENALTY FOR MURDER Dependent	.055	.011		.000[d]
		RACE OF RESPONDENT ...	.033	.007		.000[d]

a. Not assuming the null hypothesis.

b. Using the asymptotic standard error assuming the null hypothesis.

c. Cannot be computed because the asymptotic standard error equals zero.

d. Based on chi-square approximation

Typically, lambda is presented as an asymmetrical measure of association; this is the case in *Social Statistics for a Diverse Society* (Frankfort-Nachmias & Leon-Guerrero, 2009) as well as in *Adventures in Social Research* (Babbie et al., 2010). Given that, the value of lambda to be used can be found in the "Value" column in the row indicating the correct dependent variable. In this case, "cappun" (favor or oppose death penalty for murder) is the appropriate dependent variable. We see that lambda is 0.030, and that it is *not* statistically significant ($p = 0.247$). The p refers to the probability that the result is due to chance; a smaller number ($p = 0.05$ or less) would indicate statistical significance.

Gamma (γ), Kendall's Tau-*b*, and Somers' *d*

Gamma, Kendall's tau-*b*, and Somers' *d* are all measures of association for ordinal and dichotomous nominal variables. All three of these PRE statistics can take on values ranging from −1 to +1. A value of +1 indicates that there is a deterministic and positive association, such that all of the variation in the dependent variable is accounted for by the variation in the independent variable. A value of −1 indicates, again, that there is a deterministic association, but that it is negative. Although all of the variation in the dependent variable is accounted for by the variation in the independent variable, the association is in the opposite (negative) direction. When gamma, Kendall's tau-*b*, or Somers' *d* is equal to 0, this is an indication that there is no association; none of the variation in the dependent variable can be explained by the variation in the independent variable. Of course, the closer the value of either of these measures is to zero, the weaker the association. The closer the value is to either +1 or −1, the stronger the association, in the respective direction.

Each of these three measures is calculated using the concept of pairs. If all possible dyads were selected from a data set and we looked at how each value in the pair "scored" on each of two variables, we could see whether they both score higher on one than the other or whether one scores higher on one variable and lower on the other. In the former case, the pairs are called *same order pairs* and denoted *Ns*, whereas pairs in the latter case are labeled *inverse ordered pairs* and denoted *Nd*. We could also see if the two values in each pair have the same score on one or both variables (independent variable *X* or dependent variable *Y*). Those with equal values (tied) on the independent variables are represented by *Tx*, whereas those tied on the dependent variable are marked with *Ty*.

Gamma (γ), sometimes referred to as Goodman and Kruskal's gamma, is a symmetrical measure of association that is calculated as follows:

$$\gamma = \frac{Ns - Nd}{Ns + Nd}$$

Kendall's tau-*b* is also a symmetrical measure of association. The following equation shows how Kendall's tau-*b* is calculated using pairs:

$$tau\text{-}b = \frac{Ns - Nd}{\sqrt{(Ns + Nd + Tx)(Ns + Nd + Ty)}}$$

Somers' *d* is an asymmetrical measure of association; SPSS Statistics uses the term "directional measure" to describe asymmetrical measures. Somers' *d* is calculated as follows:

$$d = \frac{Ns - Nd}{Ns + Nd + Ty}$$

To compute measures of association like gamma, Kendall's tau-*b*, or Somers' *d*, use the following guidelines:

Analyze → Descriptive Statistics → Crosstabs . . .

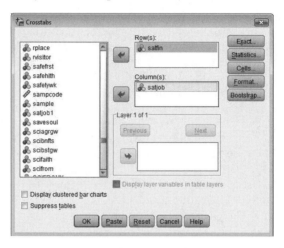

You will be given a "Crosstabs" dialog box. For this example, select "satfin" as the row variable and "satjob" as the column variable. "Satfin" is the variable representing how satisfied the respondent was with his or her financial situation. "Satjob" reveals the level of satisfaction that the respondent feels about his or her job or housework.

Now, click the "Cells" button. The following dialog box will appear:

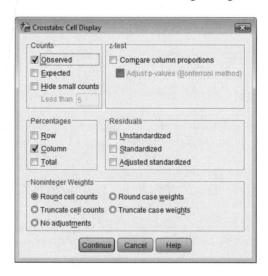

In the "Cell Display" dialog box, make sure that "Observed" counts are selected and that "Column" percentages have been requested. Now, click

"Continue." You will be returned to the "Crosstabs" dialog box. Here, click the "Statistics" button. You will be given the following dialog box:

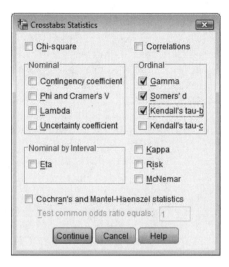

In this box, select gamma, Somers' *d*, and Kendall's tau-*b*. Click "Continue," and then click "OK" once you are returned to the original "Crosstabs" dialog box. The tables below come from the output that SPSS Statistics will create.

SATISFACTION WITH FINANCIAL SITUATION * JOB OR HOUSEWORK Crosstabulation

			JOB OR HOUSEWORK				
			VERY SATISFIED	MOD. SATISFIED	A LITTLE DISSAT	VERY DISSATISFIED	Total
SATISFACTION WITH FINANCIAL SITUATION	SATISFIED	Count	200	98	10	2	310
		% within JOB OR HOUSEWORK	26.9%	17.3%	6.5%	3.2%	20.3%
	MORE OR LESS	Count	353	268	64	21	706
		% within JOB OR HOUSEWORK	47.4%	47.4%	41.3%	33.9%	46.3%
	NOT AT ALL SAT	Count	191	199	81	39	510
		% within JOB OR HOUSEWORK	25.7%	35.2%	52.3%	62.9%	33.4%
Total		Count	744	565	155	62	1526
		% within JOB OR HOUSEWORK	100.0%	100.0%	100.0%	100.0%	100.0%

Note that the standard cross-tabulation is produced above and gives an overview by column percents of the relationship between the two variables.

Directional Measures

			Value	Asymp. Std. Error[a]	Approx. T[b]	Approx. Sig.
Ordinal by Ordinal	Somers' d	Symmetric	.208	.022	9.347	.000
		SATISFACTION WITH FINANCIAL SITUATION Dependent	.211	.022	9.347	.000
		JOB OR HOUSEWORK Dependent	.204	.022	9.347	.000

a. Not assuming the null hypothesis.

b. Using the asymptotic standard error assuming the null hypothesis.

The value for Somers' d is located in the "Value" column in the row with the appropriate variable listed as the dependent variable. (Note that because Somers' d is asymmetrical, the two values given, where the dependent variables are different, turn out to be different.) Somers' d is statistically significant in this case ($p = 0.000$) regardless of which variable is treated as the dependent variable. Its value is 0.211 with "satfin" treated as the dependent variable. We would interpret this by saying that 21.1% of the variation in satisfaction with financial situation can be accounted for by the variation in satisfaction with job or housework. Somers' d would be 0.204 with "satjob" as the dependent variable. This tells us that 20.4% of the variation in job satisfaction can be explained by the variation in satisfaction with financial situation. Either way you look at it, errors of prediction are reduced by about one-fifth.

Symmetric Measures

		Value	Asymp. Std. Error[a]	Approx. T[b]	Approx. Sig.
Ordinal by Ordinal	Kendall's tau-b	.208	.022	9.347	.000
	Gamma	.332	.034	9.347	.000
N of Valid Cases		1526			

a. Not assuming the null hypothesis.

b. Using the asymptotic standard error assuming the null hypothesis.

Above, Kendall's tau-b is given as 0.208. It is also statistically significant, where $p = 0.000$. When Kendall's tau-$b = 0.208$, we can say that 20.8% of the variation in the dependent variable can be accounted for by the variation in the independent variable.

Note the value for gamma: 0.332. It, too, is statistically significant ($p = 0.000$). When $\gamma = 0.332$, we can say that 33.2% of the variation in the dependent variable can be explained by the variation in the independent variable. It is important to determine which measure best suits the variables under analysis so that the appropriate proportion of reduction in error can be computed.

References

Babbie, E., Halley, F., Wagner, W. E., III, & Zaino, J. (2010). *Adventures in social research: Data analysis using IBM® SPSS® Statistics* (7th ed.). Thousand Oaks, CA: Pine Forge.

Frankfort-Nachmias, C., & Leon-Guerrero, A. (2009). *Social statistics for a diverse society* (5th ed.). Thousand Oaks, CA: Pine Forge.

7

Correlation and Regression Analysis

Regression analysis allows us to predict one variable from information that we have about other variables. In this chapter, linear regression will be discussed. Linear regression is a type of analysis that is performed on interval and ratio variables (labeled "scale" variables in SPSS Statistics). However, it is possible to incorporate data from variables with lower levels of measurement (i.e., nominal and ordinal variables) through the use of dummy variables. We will begin with a bivariate regression example and then add some more detail to the analysis.

Bivariate Regression

In the case of bivariate regression, researchers are interested in predicting the value of the dependent variable, Y, from the information they have about the independent variable, X. We will use the example below, where respondent's occupational prestige score is predicted from number of years of education. Choose the following menus to begin the bivariate regression analysis:

Analyze → Regression → Linear . . .

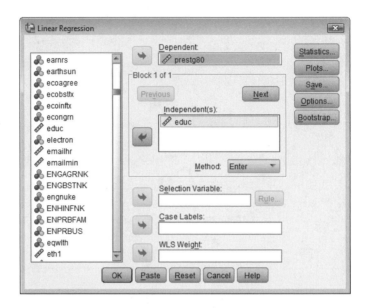

The "Linear Regression" dialog box will appear. Initially, select the variables of interest and drag them into the appropriate areas for dependent and independent variables. The variable "prestg80," respondent's occupational prestige score, should be moved to the "Dependent" area, and "educ," respondent's number of years of education, should be moved to the "Independent" area. Now, simply click "OK." The following SPSS Statistics output will be produced.

Model Summary

Model	R	R Square	Adjusted R Square	Std. Error of the Estimate
1	.516[a]	.266	.266	11.796

a. Predictors: (Constant), HIGHEST YEAR OF SCHOOL COMPLETED

In the first column of the Model Summary, the output will yield Pearson's r (in the column labeled "R"), followed in the next column by r-square (r^2). SPSS Statistics also computes an adjusted r^2 for those interested in using that value. R-square, like lambda, gamma, Kendall's tau-b, and Somers' d, is a PRE (proportional reduction in error) statistic that reveals the proportional reduction in error by introducing the dependent variable(s). In this case, $r^2 = 0.266$, which means that 26.6% of the variation in occupational prestige score is explained by the variation in years of education.

ANOVA[a]

Model		Sum of Squares	df	Mean Square	F	Sig.
1	Regression	95191.792	1	95191.792	684.139	.000[b]
	Residual	262280.857	1885	139.141		
	Total	357472.649	1886			

a. Dependent Variable: RS OCCUPATIONAL PRESTIGE SCORE (1980)

ANOVA (analysis of variance) values, including the F statistic, are given in the above table of the linear regression output.

Coefficients[a]

Model		Unstandardized Coefficients		Standardized Coefficients	t	Sig.
		B	Std. Error	Beta		
1	(Constant)	12.457	1.219		10.218	.000
	HIGHEST YEAR OF SCHOOL COMPLETED	2.290	.088	.516	26.156	.000

a. Dependent Variable: RS OCCUPATIONAL PRESTIGE SCORE (1980)

The coefficients table reveals the actual regression coefficients for the regression equation, as well as their statistical significance. In the "Unstandardized Coefficients" column, and in the "B" sub-column, the coefficients are given. In this case, the "b" value for number of years of education completed is 2.290. The "a" value, or constant, is 12.457. By looking in the last column (Sig.), you can see that both values are statistically significant ($p = 0.000$). Remember, the p value refers to the probability that the result is due to chance; so, smaller numbers are better. The standard in social sciences is usually 0.05; something is deemed statistically significant if the p value is less than 0.05. We would write the regression equation describing the model computed by SPSS Statistics as follows:

$$\hat{Y} = bX + a \rightarrow \hat{Y} = 2.394X^* + 11.542^*$$

*Statistically significant at the $p \leq 0.05$ level.

The coefficient in the bivariate regression model above can be interpreted to mean that each additional year of education provides a 2.29-point predicted increase in occupational prestige score. The constant gives the predicted occupational prestige score when years of education is zero; however, as is often the case with a regression equation, that may be beyond the range of the data for reasonable prediction. In other words, if no one had zero or near-zero years of education in the sample, then range of the data upon which the prediction was calculated did not include such, and we should be cautious in making predictions at those levels.

Correlation

Information about correlation tells us the extent to which variables are related. Below, the Pearson method of computing correlation is requested through SPSS Statistics. To examine a basic correlation between two variables, use the following menus:

Analyze → Correlate → Bivariate . . .

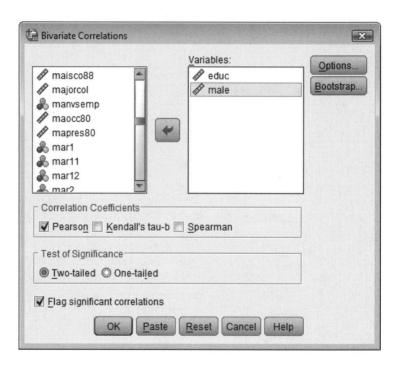

In the "Bivariate Correlations" dialog box, choose the variables that you wish to examine. In the above case, "male" (a dummy variable representing sex, described in further detail below, under "Multiple Regression") and "educ," representing years of education, have been selected. For now, "male" is a recoded version of the sex variable, where a male is coded as 1 and a female is coded as 0. Thus, a "1" indicates "male" and a "0" indicates "not male," with a proportional range in between. This allows us to treat a nominal dichotomy as an interval/ratio variable and then use it in regression and correlation analysis. The output that results is shown in the following table.

Correlations

		HIGHEST YEAR OF SCHOOL COMPLETED	Male
HIGHEST YEAR OF SCHOOL COMPLETED	Pearson Correlation	1	-.004
	Sig. (2-tailed)		.844
	N	2039	2039
Male	Pearson Correlation	-.004	1
	Sig. (2-tailed)	.844	
	N	2039	2044

Note that in the output, the correlation is an extremely low (–)0.004, which is not statistically significant ($p = 0.844$). This tells us that being male is not correlated with higher years of education.

It is also possible to produce partial correlations. Suppose you are interested in examining the correlation between occupational prestige and education. Further suppose you wish to determine the way that sex affects that correlation. Follow the following menus to produce a partial correlation:

Analyze → Correlate → Partial . . .

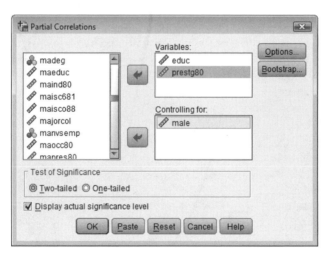

In the "Partial Correlations" dialog box, you will be able to select the variables among which you wish to examine a correlation. You will also be able to select the control variable, around which partial correlations will be computed. In this case, years of education ("educ") and occupational prestige score ("prestg80") have been selected for correlation analysis. The control variable is "male." (It is also possible to include more than one control variable.)

SPSS Statistics provides the following output.

Correlations

Control Variables			HIGHEST YEAR OF SCHOOL COMPLETED	RS OCCUPATIONAL PRESTIGE SCORE (1980)
Male	HIGHEST YEAR OF SCHOOL COMPLETED	Correlation	1.000	.516
		Significance (2-tailed)	.	.000
		df	0	1884
	RS OCCUPATIONAL PRESTIGE SCORE (1980)	Correlation	.516	1.000
		Significance (2-tailed)	.000	.
		df	1884	0

Here, the correlation is noteworthy, at 0.516, and is statistically significant ($p = 0.000$). Correlation information about variables is useful to have before constructing regression models. Should you want to know more, many textbooks in statistics and research methods have detailed discussions about how this information aids in regression analysis.

Multiple Regression

Now, suppose a researcher wished to include one or more additional independent variables in the bivariate regression analysis. This is very easy to do using SPSS Statistics. All you need to do is move the additional variables into the "Independent(s)" area in the "Linear Regression" dialog box, as seen below:

Analyze → Regression → Linear . . .

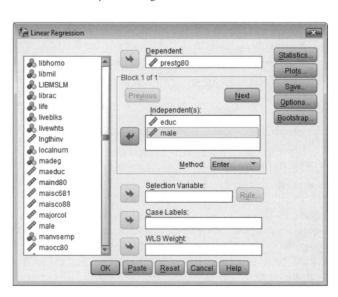

Because linear regression requires interval-ratio variables, one must take care when incorporating variables such as sex, race/ethnicity, religion, and the like. By creating dummy variables from the categories of these nominal variables, you can add this information to the regression equation.

To do this, use the recode function (for more information about recoding variables, see Chapter 2: Transforming Variables). Create a dichotomous variable for all but one category, the "omitted" comparison category/attribute, and insert each of those dichotomies into the "Independent(s)" area. The number of dummy variables necessary for a given variable will be equal to $K - 1$, where K is the number of categories of the variable. Dichotomies are an exception to the cumulative property of levels of measurement, which tells us that variables measured at higher levels can be treated at lower levels, but *not* vice versa. Dichotomies, typically considered categorical or nominal, can be "coded" to be treated as if they are at any level of measurement.

For the case of sex, we already have a dichotomy exclusive of transgender and other conditions, so the recode just changes this to one variable: "male." (Alternatively, you could have changed it to "female.") The coding should be binary: 1 for affirmation of the attribute, 0 for respondents not possessing the attribute. Now, as was entered into the previous dialog box, just select the new recoded variable, "male," from the variable bank on the left and drag it into the "Independent(s)" area on the right. You may need to set the variable property to scale in the Variable View tab of the Data Editor window, so that SPSS Statistics will allow that variable to be included in the regression analysis. Newer versions of SPSS Statistics track variable types and often will not allow you to include variables with lower levels of measurement in analyses requiring variables with higher levels of measurement.

After recoding as necessary and dragging your variables of interest into their respective areas, click the "Plots" button, and you will be given the "Plots" dialog box:

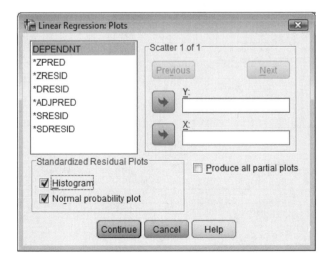

Here, you can avail yourself of a couple of useful graphics: a histogram and a normal probability plot. Click each box to request them. Then click "Continue."

When you are returned to the "Linear Regression" dialog box, now select the "Statistics" button. The following dialog box will appear:

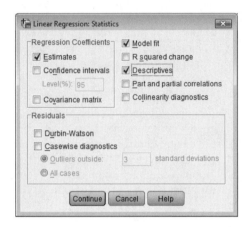

There are a number of options, including descriptive statistics, that you may select to be included in the SPSS Statistics linear regression output. For now, leave the defaults checked as shown, and click "Continue" in this box; then click "OK" when returned to the "Linear Regression" dialog box.

Below, find tables from the SPSS Statistics output that results. The first table reports the descriptive statistics that were requested. The next two tables give the same sort of information as before in the bivariate regression case: Pearson's r (correlation coefficient), r^2 (PRE), and ANOVA (analysis of variance) values.

Descriptive Statistics

	Mean	Std. Deviation	N
RS OCCUPATIONAL PRESTIGE SCORE (1980)	43.54	13.767	1887
HIGHEST YEAR OF SCHOOL COMPLETED	13.57	3.102	1887
Male	.44	.497	1887

Model Summary[b]

Model	R	R Square	Adjusted R Square	Std. Error of the Estimate
1	.516[a]	.267	.266	11.797

a. Predictors: (Constant), Male, HIGHEST YEAR OF SCHOOL COMPLETED

b. Dependent Variable: RS OCCUPATIONAL PRESTIGE SCORE (1980)

In this case, $r^2 = 0.267$, which means 26.7% of the variation in occupational prestige score ("prestg80") is explained by the variation in the independent variables: years of education ("educ") and sex ("male").

ANOVA[a]

Model		Sum of Squares	df	Mean Square	F	Sig.
1	Regression	95278.734	2	47639.367	342.314	.000[b]
	Residual	262193.915	1884	139.169		
	Total	357472.649	1886			

a. Dependent Variable: RS OCCUPATIONAL PRESTIGE SCORE (1980)

b. Predictors: (Constant), Male, HIGHEST YEAR OF SCHOOL COMPLETED

The "Coefficients" table (below), again, provides the information that can be used to construct the regression model/equation. Note that the dummy variable, "male," was not statistically significant.

$$\hat{Y} = bX_1 + bX_2 + a \rightarrow \hat{Y} = 2.29X_1^* + 0.432X_2 + 12.263^*$$

*Statistically significant at the $p \leq 0.05$ level.

The X_1 coefficient ("educ," years of education) can be interpreted to mean that each additional year of education provides a 2.29-point predicted increase in occupational prestige score. The X_2 coefficient ("male," dummy variable for sex) can be interpreted to mean that males have an occupational prestige score 0.432 points more than non-males (females). However, pay attention to the statistical significances. Although the constant and years of education are both statistically significant with $p = 0.000$, the "male" variable has a p value of 0.429, and so sex is not statistically significant in this equation.

Coefficients[a]

Model		Unstandardized Coefficients		Standardized Coefficients		
		B	Std. Error	Beta	t	Sig.
1	(Constant)	12.263	1.243		9.862	.000
	HIGHEST YEAR OF SCHOOL COMPLETED	2.290	.088	.516	26.155	.000
	Male	.432	.547	.016	.790	.429

a. Dependent Variable: RS OCCUPATIONAL PRESTIGE SCORE (1980)

The two graphics that follow show a histogram of the regression standardized residual for the dependent variable and the observed by expected cumulative probability for the dependent variable, occupational prestige.

Histogram
Dependent Variable: Rs Occupational Prestige Score (1980)

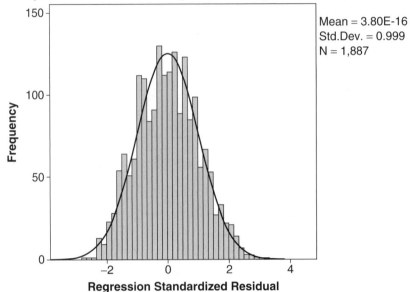

Mean = 3.80E-16
Std.Dev. = 0.999
N = 1,887

Normal P-P Plot of Regression Standardized Residual
Dependent Variable: Rs Occupational Prestige Score (1980)

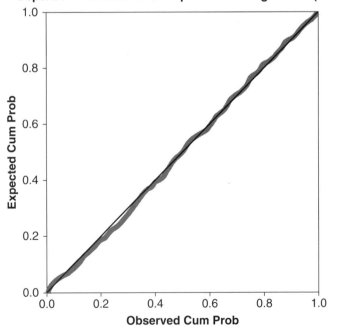

It is possible to add additional variables to your linear regression model, such as those in the dialog box featured below. Interval-ratio variables may be included, as well as dummy variables, along with others such as interaction variables. Interaction variables may be computed using the compute function (in the "Transform" menu). More information about computing variables can be found in Chapter 2: Transforming Variables. The computation would consist of variable1*variable2 = interaction variable.

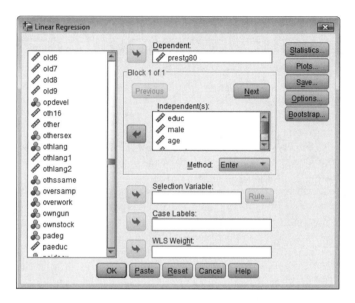

8

Logistic Regression Analysis

Much like ordinary least squares (OLS) linear regression analysis (see Chapter 7: Correlation and Regression Analysis), logistic regression analysis allows us to predict values on a dependent variable from information that we have about other (independent) variables. Logistic regression analysis is also known as *logit* regression analysis, and it is performed on a dichotomous dependent variable and dichotomous independent variables. Through the use of dummy variables, it is possible to incorporate independent variables that have more than two categories. The dependent variable usually measures the presence of something or the likelihood that a future event will happen; examples include predictions of whether students will graduate on time or whether a student will graduate at all.

Preparing Variables for Use in Logistic Regression Analysis

In order to be able to compute a logistic regression model with SPSS Statistics, all of the variables to be used should be dichotomous. Furthermore, they should be coded as "1," representing existence of an attribute, and "0" to denote none of that attribute. This may involve considerable recoding, even from dichotomies between "1" and "2" to dichotomies between "0" and "1."

For our example logistic regression, suppose we were interested in work status as predicted by gender and race/ethnicity. In Chapter 7: Correlation and Regression Analysis, "male," a dummy variable for sex, was created; the "1–2" dichotomy was transformed into a "0–1" dichotomy, where 0 = female and 1 = male.

The GSS variable "wrkstat" has eight categories, so we will need to decide what one category we want to examine. In recoding this variable, we create a variable that shows whether the respondent is (a) working full-time, or (b) not working full-time. To carry out this transformation, follow the instructions below:

Transform → Recode into Different Variables . . .

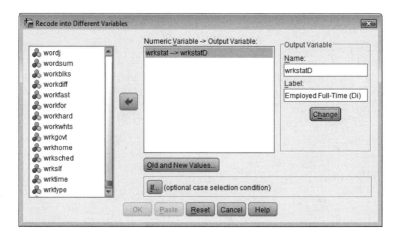

In the dialog box that appears, find the original variable, "wrkstat," and drag it into the "Numeric Variable → Output Variable" area. Now, choose a new name (the example here is "wrkstatD") and label [Employed Full-Time(Di)]. Click the "Change" button, and verify that the new variable name change has been marked next to the original variable name. Now click the "Old and New Values . . ." button.

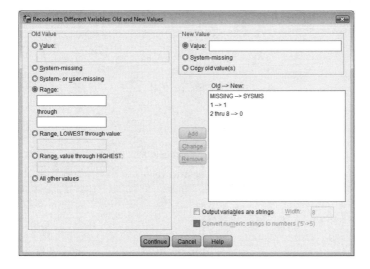

In the "Old and New Values" dialog box, have the old value of 1 (employed full-time) coded as 1 in the new variable. All other values, or those from 2 to 8, should be coded as 0 because they represent individuals who are not employed full-time. Be sure to include an instruction to SPSS/PASW Statistics for missing cases, such as the one in the screen image above.

Now, click "Continue" and then "OK" in the "Recode into Different Variables" box, and the new variable will be created. You will want to update information about "wrkstatD" in the Variable View tab of the Data Editor window, such as its measure (nominal), decimals (0), and so forth, as shown below:

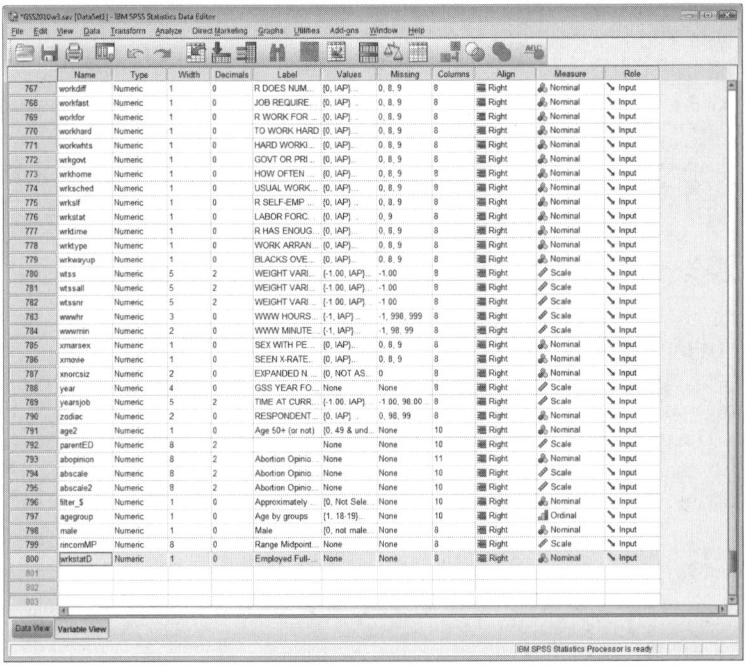

	Name	Type	Width	Decimals	Label	Values	Missing	Columns	Align	Measure	Role
767	workdiff	Numeric	1	0	R DOES NUM...	{0, IAP}...	0, 8, 9	8	Right	Nominal	Input
768	workfast	Numeric	1	0	JOB REQUIRE...	{0, IAP}...	0, 8, 9	8	Right	Nominal	Input
769	workfor	Numeric	1	0	R WORK FOR ...	{0, IAP}...	0, 8, 9	8	Right	Nominal	Input
770	workhard	Numeric	1	0	TO WORK HARD	{0, IAP}...	0, 8, 9	8	Right	Nominal	Input
771	workwhts	Numeric	1	0	HARD WORKI...	{0, IAP}...	0, 8, 9	8	Right	Nominal	Input
772	wrkgovt	Numeric	1	0	GOVT OR PRI...	{0, IAP}...	0, 8, 9	8	Right	Nominal	Input
773	wrkhome	Numeric	1	0	HOW OFTEN ...	{0, IAP}...	0, 8, 9	8	Right	Nominal	Input
774	wrksched	Numeric	1	0	USUAL WORK...	{0, IAP}...	0, 8, 9	8	Right	Nominal	Input
775	wrkslf	Numeric	1	0	R SELF-EMP ...	{0, IAP}...	0, 8, 9	8	Right	Nominal	Input
776	wrkstat	Numeric	1	0	LABOR FORC...	{0, IAP}...	0, 9	8	Right	Nominal	Input
777	wrktime	Numeric	1	0	R HAS ENOUG...	{0, IAP}...	0, 8, 9	8	Right	Nominal	Input
778	wrktype	Numeric	1	0	WORK ARRAN...	{0, IAP}...	0, 8, 9	8	Right	Nominal	Input
779	wrkwayup	Numeric	1	0	BLACKS OVE...	{0, IAP}...	0, 8, 9	8	Right	Nominal	Input
780	wtss	Numeric	5	2	WEIGHT VARI...	{-1.00, IAP}...	-1.00	8	Right	Scale	Input
781	wtssall	Numeric	5	2	WEIGHT VARI...	{-1.00, IAP}...	-1.00	8	Right	Scale	Input
782	wtssnr	Numeric	5	2	WEIGHT VARI...	{-1.00, IAP}...	-1.00	8	Right	Scale	Input
783	wwwhr	Numeric	3	0	WWW HOURS...	{-1, IAP}...	-1, 998, 999	8	Right	Scale	Input
784	wwwmin	Numeric	2	0	WWW MINUTE...	{-1, IAP}...	-1, 98, 99	8	Right	Scale	Input
785	xmarsex	Numeric	1	0	SEX WITH PE...	{0, IAP}...	0, 8, 9	8	Right	Nominal	Input
786	xmovie	Numeric	1	0	SEEN X-RATE...	{0, IAP}...	0, 8, 9	8	Right	Nominal	Input
787	xnorcsiz	Numeric	2	0	EXPANDED N...	{0, NOT AS...	0	8	Right	Nominal	Input
788	year	Numeric	4	0	GSS YEAR FO...	None	None	8	Right	Scale	Input
789	yearsjob	Numeric	5	2	TIME AT CURR...	{-1.00, IAP}...	-1.00, 98.00 ...	8	Right	Scale	Input
790	zodiac	Numeric	2	0	RESPONDENT...	{0, IAP}...	0, 98, 99	8	Right	Nominal	Input
791	age2	Numeric	1	0	Age 50+ (or not)	{0, 49 & und...	None	10	Right	Nominal	Input
792	parentED	Numeric	8	2		None	None	10	Right	Scale	Input
793	abopinion	Numeric	8	2	Abortion Opinio...	None	None	11	Right	Nominal	Input
794	abscale	Numeric	8	2	Abortion Opinio...	None	None	10	Right	Scale	Input
795	abscale2	Numeric	8	2	Abortion Opinio...	None	None	10	Right	Scale	Input
796	filter_$	Numeric	1	0	Approximately ...	{0, Not Sele...	None	10	Right	Nominal	Input
797	agegroup	Numeric	1	0	Age by groups	{1, 18-19}...	None	10	Right	Ordinal	Input
798	male	Numeric	1	0	Male	{0, not male...	None	8	Right	Nominal	Input
799	rincomMP	Numeric	8	0	Range Midpoint...	None	None	8	Right	Scale	Input
800	wrkstatD	Numeric	1	0	Employed Full-...	None	None	8	Right	Nominal	Input
801											
802											
803											

You can verify the new variable in the Data View window. You could also produce a frequency distribution of the new variable, "wrkstatD," to verify the transformation and observe the distribution, as given below:

Employed Full-Time (Di)

		Frequency	Percent	Valid Percent	Cumulative Percent
Valid	0	1124	55.0	55.1	55.1
	1	917	44.9	44.9	100.0
	Total	2041	99.9	100.0	
Missing	System	3	.1		
Total		2044	100.0		

Notice the approximately even distribution of "wrkstatD" (full-time employment dichotomy). If there were only a small percentage in either category, the variable would not be suitable for use in this analysis and would need to be recoded in some other way to capture more variation.

Creating a Set of Dummy Variables to Represent a Multicategory Nominal Variable

Our other independent variables will represent race/ethnicity. We will use the four-category race variable that was recoded using the original race variable and the variable "Hispanic." The categories are White, Black, Other, and Hispanic. Because there are four categories, we will need to create three dummy variables. The number of dummy variables in a set that represents a nominal variable is equal to $K - 1$, where K is the number of categories. To do this, first produce a frequency distribution, as follows:

Analyze → Descriptive Statistics → Frequencies . . .

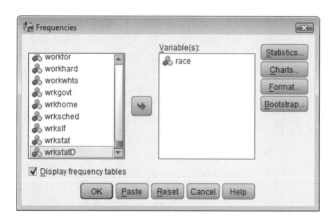

After requesting the frequency for race (as recoded with additional information from the variable "Hispanic" in Chapter 2: Transforming Variables), we get the following chart:

RACE OF RESPONDENT

		Frequency	Percent	Valid Percent	Cumulative Percent
Valid	WHITE	1423	69.6	69.6	69.6
	BLACK	299	14.6	14.6	84.2
	OTHER	87	4.3	4.3	88.5
	Hispanic	235	11.5	11.5	100.0
	Total	2044	100.0	100.0	

Because the "Other" category is relatively small at just 4.3% of valid cases, you might wish to define those in that category as missing and proceed with a three-category variable (White, Black, and Hispanic), producing two dummy variables: $K - 1 = 2$.

We will proceed with all four categories. Let's make four dummy variables, and then we'll choose which three to use for the analysis based on which groups we wish to compare. First select the following menus:

Transform → Compute Variable . . .

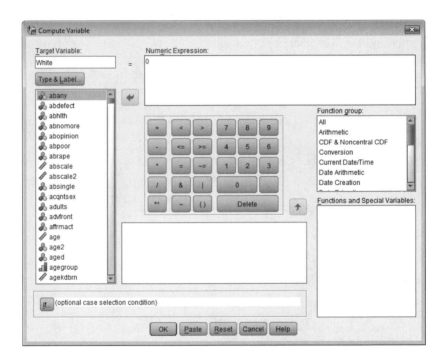

Name the target variable; in this case, it will be "White." Now, set all cases in the new variable equal to zero, as above. Then, click "OK." You can verify in the Data Editor window that a new variable has been appended to the data file that has all zeros as data.

Next, we will want to correct those cases for which the respondent was White. To do that, again select these menus:

Transform → Compute Variable . . .

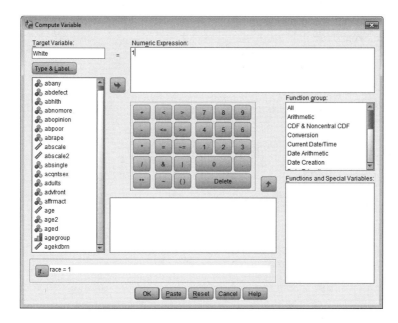

Now, change the value in the Numeric Expression area to 1. Then, click "If . . ." in the lower left corner of the "Compute Variable" box. The following dialog box will appear:

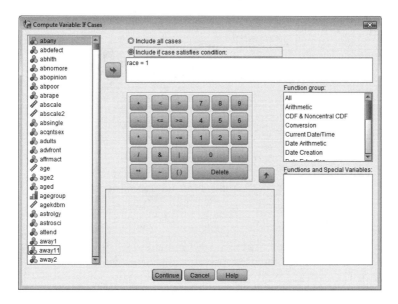

Select the radio button next to "Include if case satisfies condition:". Then type "race = 1" in the box beneath. Now click "Continue" and then "OK." You'll be given the following warning. Be sure to click "OK" to verify that you do wish to alter some of the data in this variable.

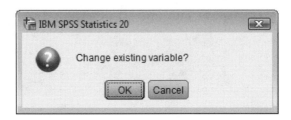

Now, let's use the same procedure to create the variable "Black." Click the following menus:

Transform → Compute Variable . . .

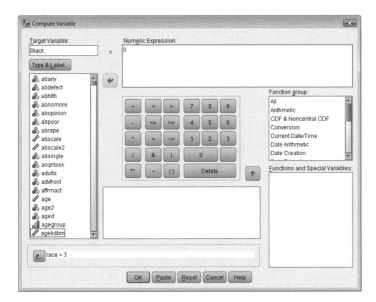

First, click the "Reset" button to eliminate information from the prior transformation. Then set all of the cases in the new variable, "Black," equal to zero. Then click "OK." Once you verify the creation of the new variable in your Data Editor window, again click these menus:

Transform → Compute Variable . . .

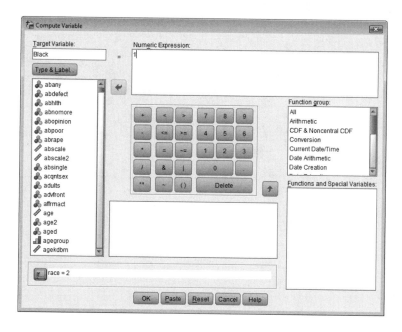

Here, in the Numeric Expression area, you will enter 1. Now, click the "If . . ." button, and in the dialog box that appears, choose the "Include if case satisfies condition:" radio button and enter the equation "race = 2" in the box beneath. Now, click "Continue" and then "OK."

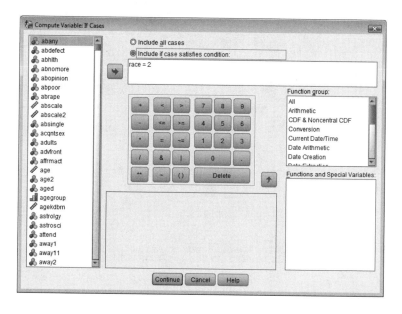

Next, follow the same procedure for "Other/Others" and "Hispanic/ Latino," as described below:

1. For "Others," select Transform → Compute Variable . . . , and then set Others = 0. Next, using the Compute command, set Others = 1, IF: race = 3. (We are using the name "Others" instead of "Other," since "Other" represents another variable in the GSS original data file. You are free to label this new variable as "Other," though that would delete the original variable that contains data about other Protestant denominations.)

2. For "Latino," select Transform → Compute Variable . . . , then set Latino = 0. Next, using the Compute command, set Latino = 1, IF: race = 4. (We are using the name "Latino" because "Hispanic" represents another variable in the GSS original data file. In fact, the original variable, "Hispanic," contains information that we used to recode the very race variable we are using here by adding a fourth category in Chapter 2: Transforming Variables. You could rename "Hispanic" to something else and use that name here if you wish.)

You could also accomplish the steps above by using the syntax editor (see Chapter 12: Advanced Applications). When familiar with this method, it can save a great deal of time, particularly when dealing with larger numbers of variables/categories.

File → New → Syntax . . .

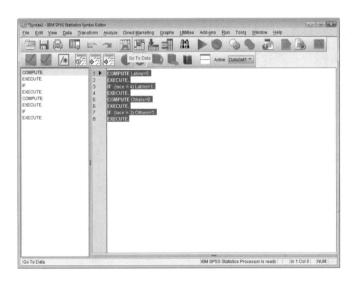

Once you type and select the above SPSS Statistics command syntax, click the green "play" triangle button and SPSS Statistics will carry out the commands and create the new variables as desired.

Now, for all four new variables, go to the Variable View window and edit the settings to reflect each new variable's decimals (0), measure (nominal), and so on.

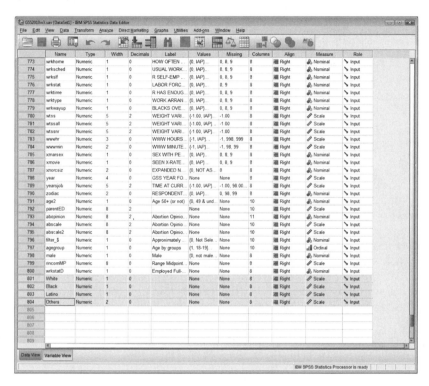

The method we have used above creates dummy variables in two steps: (a) Name the new variable and set all cases equal to zero, and (b) change the settings for the appropriate cases corresponding to the race/ethnicity equal to one. Note that we are able to set the variables equal to zero first, without worrying about missing cases, only because there happen to be no missing cases for the race variable. If there were missing cases, we would need to handle those cases with the Compute command as well.

Logistic Regression Analysis

Choose the following menus to begin the logistic regression analysis:

Analyze → Regression → Binary Logistic . . .

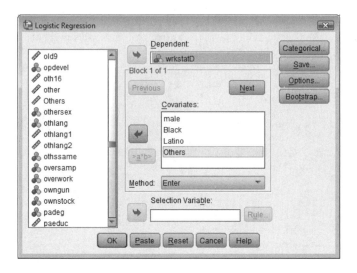

Select "wrkstatD" as the dependent variable by dragging it from the column on the left into the box under "Dependent." Independent variables will go in the box labeled "Covariates." For gender, use "male." For race/ethnicity, think about which category you want to use as a reference. Whatever results you get will reveal whether other groups are more or less likely to have full-time employment than the reference (left out) category.

In this case, if we omit "White," then the comparison can be made from other groups to Whites. Let's use that in our example. Enter the three other dummy variables as they appear in the screen image above. Then click "OK." A selection from the output produced is as follows. See "Interpreting Odds Ratios" at the end of this chapter for an explanation of how to interpret the central parts of the logistic regression output.

Model Summary

Step	-2 Log likelihood	Cox & Snell R Square	Nagelkerke R Square
1	2772.750[a]	.017	.023

a. Estimation terminated at iteration number 3 because parameter estimates changed by less than .001.

Variables in the Equation

		B	S.E.	Wald	df	Sig.	Exp(B)
Step 1[a]	male	.504	.090	31.078	1	.000	1.655
	Black	-.240	.131	3.335	1	.068	.787
	Latino	.016	.142	.012	1	.912	1.016
	Others	-.033	.224	.021	1	.884	.968
	Constant	-.392	.068	33.512	1	.000	.676

a. Variable(s) entered on step 1: male, Black, Latino, Others.

Logistic Regression Using a Categorical Covariate Without Dummy Variables

The logistic regression command has a built-in way to analyze a nominal/categorical variable like our recoded race variable. The results produced will be identical to those described earlier in this chapter, and there is no need to create dummy variables. There are, however, some situations in SPSS/PASW Statistics where you must create and use dummy variables, and that method directly exposes the user to how the data are being treated and analyzed. The following method should be used only by those who fully understand the nature of categorical analysis within a logistic regression model, and it should be noted that there are limitations: The reference category can be only the first or last category of the variable to be used as a covariate. In the case of the race variable, "White" is the first category and "Latino" is the last category. Had we wished to use "Black" or "Others" as a reference category, we would need to have created dummy variables as above or to have recoded "race" to suit our needs. Another limitation will be addressed toward the end of this chapter; it concerns complications when using certain methods of inclusion for variables.

In any case, follow the instructions below to produce the logistic regression equation using the categorical variable method:

Analyze → Regression → Binary Logistic . . .

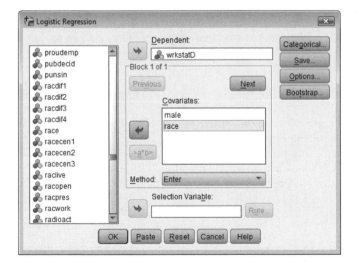

As before, select "wrkstatD" as the dependent variable. Also, add "male" as one of the covariates. Now, simply add "race" as one of the covariates. You cannot stop here and run the analysis; the results would be without useful interpretation. So it is necessary to tell SPSS Statistics that "race" is a categorical/nominal variable and to tell it which category should be the designated reference category (and therefore "left out" of the analysis in the same way it was in our utilization of dummy variables). To do this, click the "Categorical" button . . .

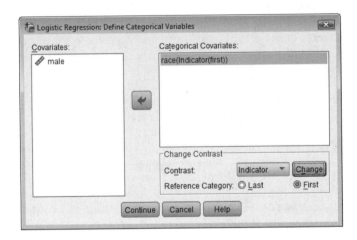

Now, in the dialog box with which you are presented, select "race" from the list of covariates on the left, and drag it into the box under "Categorical Covariates." Then select "First" as the reference category, and click the "Change" button. You should notice the word "first" appear as in the screen image above. Now click "Continue" and then "OK." The output from this command is displayed and interpreted in the following section.

Interpreting Odds Ratios

Logistic regression uses natural logarithms to produce a logistic curve as a predictor, whereas you may remember that OLS linear regression uses the least squares method to produce a straight line as a predictor. The coefficients in a logistic regression model can be exponentiated as log odds ratios. Selected output from the logistic regression command, above, has been printed below:

Model Summary

Step	-2 Log likelihood	Cox & Snell R Square	Nagelkerke R Square
1	2772.750[a]	.017	.023

a. Estimation terminated at iteration number 3 because parameter estimates changed by less than .001.

Variables in the Equation

		B	S.E.	Wald	df	Sig.	Exp(B)
Step 1[a]	male	.504	.090	31.078	1	.000	1.655
	race			3.494	3	.322	
	race(1)	-.240	.131	3.335	1	.068	.787
	race(2)	-.033	.224	.021	1	.884	.968
	race(3)	.016	.142	.012	1	.912	1.016
	Constant	-.392	.068	33.512	1	.000	.676

a. Variable(s) entered on step 1: male, race.

Notice that in the Model Summary box, two different r-squares (r^2) are presented: Cox and Snell as well as Nagelkerke. Although computed differently, these numbers can be interpreted in much the same way as r^2 itself, the coefficient of determination. (See Chapter 7: Correlation and Regression Analysis, for more details on proportional reduction in error [PRE] statistics and the coefficient of determination.)

In the "Variables in the Equation" box, the coefficients themselves are found in column "B," but they have been exponentiated in the column "Exp(B)." This value tells how much more or less likely a subject in the designated category is to be in the affirmative category on the dependent variable (employed full-time) than a subject in the omitted reference category. For the coefficient male, the odds ratio is 1.655, and it is statistically significant ($p = 0.000$ in the "Sig." column); therefore, men are 1.655 times more likely than women to be employed full-time and not fall into some other category of employment.

The first race row in the "Variables in the Equation" box is the omitted reference category, White. Notice that there is no coefficient or exponentiated coefficient in that row. In the rows below, "race(1)" represents the next category, Black; "race(2)" represents the third category, Other/Others; and "race(3)" represents the last category, Hispanic/Latino. Although none of the categories (or dummy variables) was statistically significant for race, Hispanic/Latino was the closest ($p = 0.068$). The exponentiated coefficient reveals that Hispanics were 0.787 times less likely than those in the reference category (Whites) to have full-time employment and not fall in some other category of the original variable: employed part-time, retired, keeping house, and so forth. Again, however, race was not statistically significant in this equation/model. Direct comparisons can be made only with the reference category.

To have SPSS Statistics help produce the equation with the best set of statistically significant variables, so that you will not need to try each combination manually, you can select a different option from the pull-down menu in the "Method" pane in the "Logistic Regression" dialog box, such as "Backward: Conditional," as demonstrated below. This method works effectively when using the dummy variable approach to logistic regression analysis, but it is not effective with the categorical variable method.

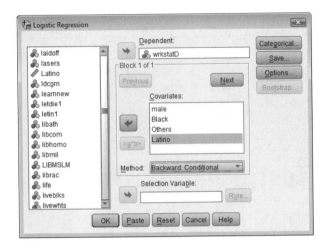

From the output produced, the following selection reveals the final coefficients. Notice that each step removes variables that are not statistically significant and that contribute the least to the model. In the third step, the final model is provided. The number of steps that are required by SPSS Statistics to produce the final equation depends on numerous factors.

Notice that in our simple example, the final equation is very similar to the model we produced above, although the nonsignificant dummy variables have been removed from the model. Notice, too, that the value of r^2 (both Cox and Snell, and Nagelkerke) is not reduced as a result of the removal of variables (i.e., in conjunction with the other variables in the equation, the variation in the deleted variables did not account for any variation in the dependent variable).

Model Summary

Step	-2 Log likelihood	Cox & Snell R Square	Nagelkerke R Square
1	2772.750[a]	.017	.023
2	2772.763[a]	.017	.023
3	2772.787[a]	.017	.023

a. Estimation terminated at iteration number 3 because parameter estimates changed by less than .001.

Variables in the Equation

		B	S.E.	Wald	df	Sig.	Exp(B)
Step 1[a]	male	.504	.090	31.078	1	.000	1.655
	Black	-.240	.131	3.335	1	.068	.787
	Others	-.033	.224	.021	1	.884	.968
	Latino	.016	.142	.012	1	.912	1.016
	Constant	-.392	.068	33.512	1	.000	.676
Step 2[a]	male	.504	.090	31.066	1	.000	1.655
	Black	-.242	.130	3.481	1	.062	.785
	Others	-.035	.223	.024	1	.876	.966
	Constant	-.390	.064	36.623	1	.000	.677
Step 3[a]	male	.504	.090	31.082	1	.000	1.655
	Black	-.240	.129	3.457	1	.063	.787
	Constant	-.392	.063	38.149	1	.000	.676

a. Variable(s) entered on step 1: male, Black, Others, Latino.

9

Testing Hypotheses Using Means and Cross-Tabulation

S PSS Statistics allows for automatic testing of hypotheses without having to make a computation and check a cut point in a table in the back of a statistics book. The actual statistical significance is presented with the results.

Comparing Means

This section explains how to examine differences between two means. Comparing means between groups requires having a variable that will allow for a division into the appropriate groups, in the same way that it is required to split a data file for a comparative analysis. (See Chapter 6: Cross-Tabulation and Measures of Association for Nominal and Ordinal Variables for further details on using the Split File command.)

Suppose you are interested in comparing the occupational prestige scores of respondents and wish to examine the differences between men and women. Select the following menus:

Analyze → Compare Means → Means . . .

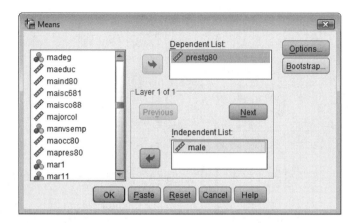

Additional layers can be requested by clicking "Next" and then adding the additional variable(s). The Data → Split File option can also be used to compare groups across categories/attributes of a variable. If you click on the "Options" button, the box shown below will appear, and you can customize the cell statistics that SPSS Statistics will report in the output.

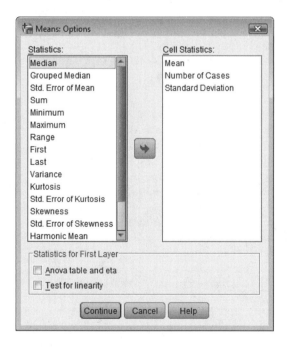

Select those statistics that you wish to be reported, and drag them to the Cell Statistics area. After you click "Continue" and then "OK" in the "Means" dialog box, SPSS Statistics will generate the following output.

Report

RS OCCUPATIONAL PRESTIGE SCORE (1980)

Male	Mean	N	Std. Deviation
not male (female)	43.36	1054	13.943
male	43.74	836	13.547
Total	43.53	1890	13.767

A simple table is provided that displays mean, sample size, and standard deviation for the total sample, as well as for each category of the variable of interest: occupational prestige.

Comparing Means: Paired-Samples *t* Test

The paired-samples *t* test can be used when, for a given data set, the means of all the values for each of two variables are to be compared. An example of such a data set would be a data file containing scores for a group of people who have taken a particular pretest and then an identical posttest after some sort of stimulus had been administered. There are many situations where this method of analysis is appropriate.

From the 2010 General Social Survey (GSS), we can look at parental education by parent's gender. In other words, we can compare the years of education completed by the respondents' fathers with those completed by the mothers. To begin, select the following menus:

Analyze → Compare Means → Paired Samples T Test . . .

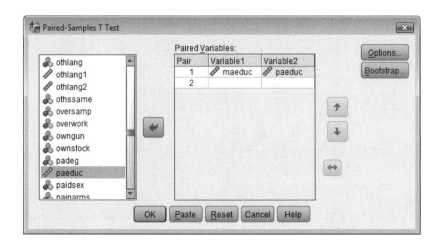

In order to move a complete entry into the Paired Variables area, you will need to select two variables from the variable bank on the left side of the dialog box. Once you have selected two variables and clicked the arrow for each (one right after the other), their names will appear under "Variable1" and "Variable2" for "Pair 1." Additional pairs may be added in subsequent rows.

Should you wish to change the confidence interval (note that the default is 95%), or change the way that missing cases are handled or excluded, click the "Options" button, and you will be given a dialog box to make those selections, as illustrated below.

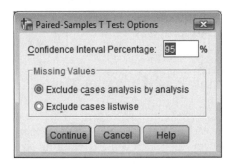

Once you have made your selections, click "Continue" in the "Options" box, and then click "OK" in the "Paired-Samples T Test" box. What follows is output that SPSS/PASW Statistics will produce to fulfill your request:

Paired Samples Statistics

		Mean	N	Std. Deviation	Std. Error Mean
Pair 1	HIGHEST YEAR SCHOOL COMPLETED, MOTHER	11.67	1406	3.730	.099
	HIGHEST YEAR SCHOOL COMPLETED, FATHER	11.63	1406	4.186	.112

Paired Samples Correlations

		N	Correlation	Sig.
Pair 1	HIGHEST YEAR SCHOOL COMPLETED, MOTHER & HIGHEST YEAR SCHOOL COMPLETED, FATHER	1406	.707	.000

The two tables above give basic information, such as mean, sample size, standard deviation, standard error, and correlation between the two variables. Note that there is a statistically significant correlation between these two variables.

Paired Samples Test

| | | Paired Differences | | | | | | | |
| | | | | | 95% Confidence Interval of the Difference | | | | |
		Mean	Std. Deviation	Std. Error Mean	Lower	Upper	t	df	Sig. (2-tailed)
Pair 1	HIGHEST YEAR SCHOOL COMPLETED, MOTHER - HIGHEST YEAR SCHOOL COMPLETED, FATHER	.033	3.059	.082	-.127	.193	.401	1405	.688

In the third table, results of the *t* test are displayed. Here, $t = 0.401$ and is not statistically significant at the 0.05 level or better.

Comparing Means: Independent-Samples *t* Test

Independent-samples *t* tests allow us to compare the mean of a particular variable across independent groups. To look at an example using occupational prestige scale scores, use the following menu selections:

Analyze → Compare Means → Independent-Samples T Test . . .

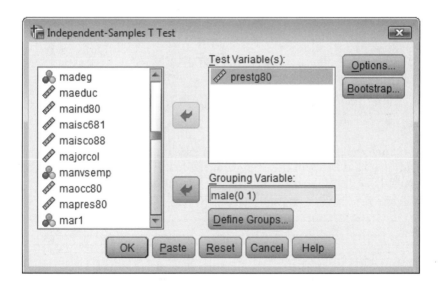

In this dialog box, drag the variable of interest, occupational prestige score ("prestg80"), into the Test Variable(s) area. Next, you will need to select the grouping variable. In the example above, "male" (dummy variable for gender) has been selected. Now, you need to inform SPSS Statistics which groups are to be compared. Click the "Define Groups" button.

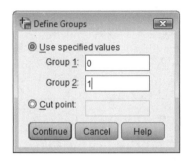

In this "Define Groups" dialog box, fill in the category values for each group. Note that depending upon how the grouping variable is categorized, you have the option of selecting a cut point to define the groups. This could be done with age, test scores, and so on.

Click "Continue," which will take you back to the "Independent-Samples T Test" dialog box. You have the option of changing the confidence interval and method of case exclusion. To make those choices, click the "Options" button and you will be given the dialog box that follows:

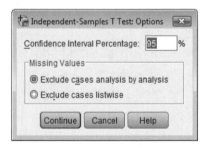

Enter the changes you would like to make, if any, then click "Continue," and then click "OK" in the original dialog box. The output below will be generated by SPSS Statistics in response to your query:

Group Statistics

Male		N	Mean	Std. Deviation	Std. Error Mean
RS OCCUPATIONAL PRESTIGE SCORE (1980)	not male (female)	1054	43.36	13.943	.429
	male	836	43.74	13.547	.469

Independent Samples Test

		Levene's Test for Equality of Variances		t-test for Equality of Means						
		F	Sig.	t	df	Sig. (2-tailed)	Mean Difference	Std. Error Difference	95% Confidence Interval of the Difference Lower	Upper
RS OCCUPATIONAL PRESTIGE SCORE (1980)	Equal variances assumed	2.957	.086	-.591	1888	.555	-.377	.638	-1.627	.874
	Equal variances not assumed			-.593	1812.810	.553	-.377	.636	-1.623	.870

Basic statistics are given in the first table. The second table reveals the results and significance (or lack thereof in this case) of the *t* test. When the *t* test is statistically significant, we can say that the groups (males and females in this case) in the sample come from populations whose mean is different. Given the failure of the *t* test to achieve statistical significance in this case, we cannot make that claim.

One-Sample *t* Test

By using a one-sample *t* test, it is possible to test whether the mean of a particular variable differs from a specified value. This test will determine whether that difference is statistically significant. To run this test, follow these instructions:

Analyze → Compare Means → One-Sample T Test . . .

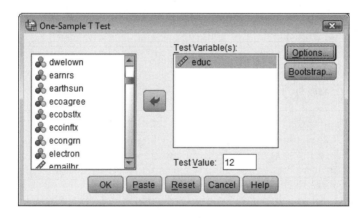

By clicking on the "Options" button you are able to select the confidence interval (default = 95%) and determine the criterion for excluding missing cases.

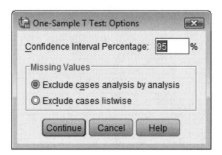

Click "Continue" then click "OK" once you are taken back to the original One-Sample T Test dialog box. The following output will be produced, demonstrating that in this case, there is a statistically significant difference between the specified value of 12 and the mean number of years of education of those in the sample.

One-Sample Statistics

	N	Mean	Std. Deviation	Std. Error Mean
HIGHEST YEAR OF SCHOOL COMPLETED	2039	13.46	3.149	.070

One-Sample Test

	Test Value = 12					
					95% Confidence Interval of the Difference	
				Mean		
	t	df	Sig. (2-tailed)	Difference	Lower	Upper
HIGHEST YEAR OF SCHOOL COMPLETED	20.948	2038	.000	1.461	1.32	1.60

Chi-Square (χ^2)

The chi-square test statistic is used to test statistical independence or goodness of fit. It is used for categorical data. Chi-square summarizes the differences between the observed frequencies (f_o) and the expected frequencies (f_e) in a bivariate table. The observed frequencies are those produced from the raw data. To compute the expected frequencies for any given cell in a bivariate table, multiply the column marginal (total) by the row marginal (total), and then divide by N (sample size). The obtained chi-square is then calculated as follows:

$$\chi^2 = \Sigma \frac{(f_o - f_e)}{f_e}.$$

There are two menu selections that will produce chi-square results: nonparametric tests and cross-tabulations. The former is demonstrated here, first using the legacy dialog (the way the function has been performed by prior versions of SPSS) and then using the new interface tools. To perform a chi-square analysis, choose the following menu options:

Analyze → Nonparametric Tests → Legacy Dialogs → Chi-Square . . .

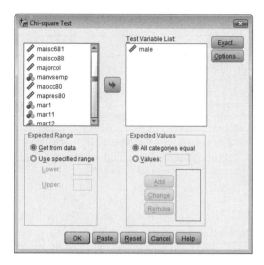

Now, in the "Chi-square Test" dialog box, choose the variable that you would like to test. In this case, we chose "male," the dummy variable representing gender. For expected values, by leaving the button selected for "All categories equal," the expectation is that 50% of the respondents will be men and 50% will be women.

If you click on the "Options" button, you will be given a dialog box that allows you to request additional information and determine the method by which cases should be excluded.

When these decisions have been made, click "Continue," and then click "OK" in the "Chi-square Test" dialog box. The output that follows will be provided by SPSS Statistics:

Male

	Observed N	Expected N	Residual
not male (female)	1153	1022.0	131.0
male	891	1022.0	-131.0
Total	2044		

Test Statistics

	Male
Chi-Square	33.583[a]
df	1
Asymp. Sig.	.000

a. 0 cells (.0%) have expected
frequencies less than 5. The
minimum expected cell frequency is
1022.0.

The first table provides the observed and expected amounts, and of course the residual or difference. The second table yields the chi-square value and degrees of freedom, and reveals whether the value is statistically significant. In this case, chi-square = 33.583 with 1 degree of freedom (df) and is statistically significant ($p = 0.000$).

The newer interface, among other things, offers greater assistance to the user in terms of selecting the appropriate statistic. It also provides a hypothesis-based response in the output, though it does not include the table of observed and expected values. Follow the menu options below to obtain chi-square results using this method:

Analyze → Nonparametric Tests → One Sample . . .

You will be presented with a dialog box like the one below. Choose the "Customize analysis" radio button, because you know that you want to calculate the chi-square statistic.

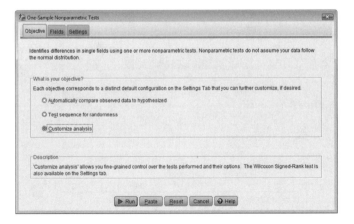

Now, click the "Fields" tab at the top of this dialog box. Here, make sure that "Use custom field assignments" has been selected and choose the variable(s) that you want to test by including them in the Test Fields area. Below, "male" has been chosen.

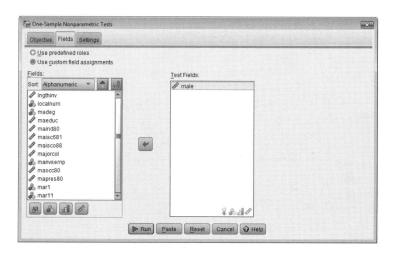

Now, click the "Settings" tab at the top of the dialog box. Click "Customize tests" and check the box for "Compare observed probabilities to hypothesized (Chi-Square test)." You can choose the "Test Options" entry on the left if you wish to change the confidence level from 95% or the significance level from 5%.

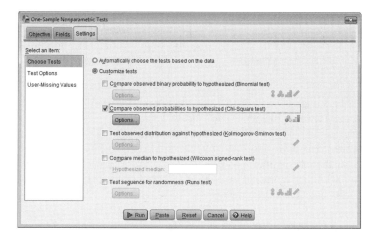

Also, you can change how missing cases are excluded under "Test Options," as well.

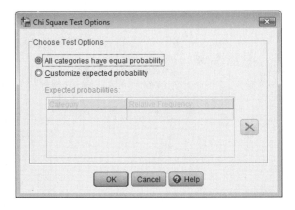

After you click "Run," SPSS Statistics will provide you with the following response.

Hypothesis Test Summary

	Null Hypothesis	Test	Sig.	Decision
1	The categories of Male occur with equal probabilities.	One-Sample Chi-Square Test	.000	Reject the null hypothesis.

Asymptotic significances are displayed. The significance level is .05.

The output clearly indicates that the null hypothesis, that males and females occur in the population with equal probabilities, should be rejected. Although there are more steps in this process than this example of the legacy dialog shows, the user has many more options and can request multiple statistical tests at once.

Chi-Square (χ^2) and Cross-Tabulation

If you would like to produce full cross-tabulations with chi-square test results, such as those detailed in Chapter 6: Cross-Tabulation and Measures of Association for Nominal and Ordinal Variables, see the following example by selecting the following menus:

Analyze → Descriptive Statistics → Crosstabs . . .

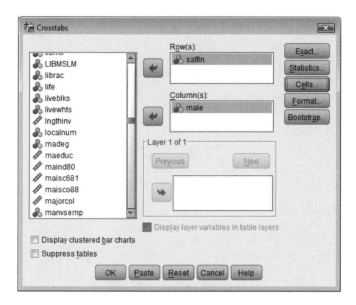

Click the "Statistics" button, and you will be presented with the following dialog box:

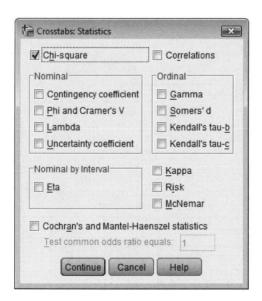

Select the "Chi-square" option, leaving a check mark in the associated box. Click "Continue" and then click the "Cells . . . " button.

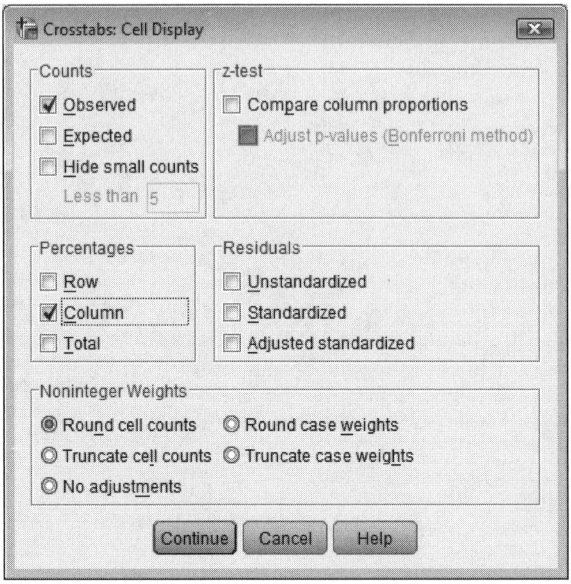

Crosstabs: Cell Display

Counts
- ☑ Observed
- ☐ Expected
- ☐ Hide small counts
  - Less than 5

z-test
- ☐ Compare column proportions
- ☐ Adjust p-values (Bonferroni method)

Percentages
- ☐ Row
- ☑ Column
- ☐ Total

Residuals
- ☐ Unstandardized
- ☐ Standardized
- ☐ Adjusted standardized

Noninteger Weights
- ◉ Round cell counts   ○ Round case weights
- ○ Truncate cell counts ○ Truncate case weights
- ○ No adjustments

[Continue] [Cancel] [Help]

After choosing "Column" percentages, click "Continue" and then "OK" back in the "Crosstabs" window. This will produce the following SPSS Statistics output:

SATISFACTION WITH FINANCIAL SITUATION * Male Crosstabulation

			Male		
			not male (female)	male	Total
SATISFACTION WITH FINANCIAL SITUATION	SATISFIED	Count	263	215	478
		% within Male	22.9%	24.2%	23.5%
	MORE OR LESS	Count	522	396	918
		% within Male	45.5%	44.5%	45.0%
	NOT AT ALL SAT	Count	363	279	642
		% within Male	31.6%	31.3%	31.5%
Total		Count	1148	890	2038
		% within Male	100.0%	100.0%	100.0%

Chi-Square Tests

	Value	df	Asymp. Sig. (2-sided)
Pearson Chi-Square	.451[a]	2	.798
Likelihood Ratio	.450	2	.798
Linear-by-Linear Association	.213	1	.644
N of Valid Cases	2038		

a. 0 cells (.0%) have expected count less than 5. The minimum expected count is 208.74.

10

Analysis of Variance

A nalysis of variance (ANOVA) is an inferential statistics technique that involves a statistical test for significance of differences between mean scores of at least two groups across one or more than one variable. ANOVA can be used to test for statistical significance using categorical independent variables in conjunction with a continuous dependent variable. ANOVA is based on the comparison of variance between groups to the variance within groups, emerging as the F ratio, or F statistic.

One-Way ANOVA

We will perform an ANOVA test using "race" as the independent variable and respondent's income as the dependent variable. To measure "race," we will use the recoded, four-category version of that variable. (For details on how to recode "race," see Chapter 2: Transforming Variables.) For income, we will need to recode "rincom06" because it is measured by numeric categories corresponding to nonequivalent ranges of annual earnings. We can use the midpoint dollar amount of each of those ranges to approximate income for respondents in each category. To make it so, follow these instructions:

Transform $\rightarrow$ Recode into Different Variables . . .

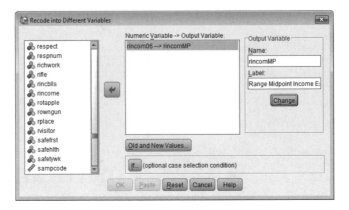

In the "Recode into Different Variables" dialog box, select "rincom06" and move it from the variable bank on the left over to the area for "Numeric Variable → Output Variable." Now, type a new name in the Output Variable area, and also a descriptive label. Next, click the "Change" button, and make sure the change was recorded in the Numeric Variable → Output Variable listing.

At this point, click the "Old and New Values" button, and you will be presented with the following dialog box:

In the "Recode into Different Variables: Old and New Values" dialog box, you will enter the numeric values from the original variable and the corresponding midpoint of each of those ranges as the new values. Don't forget to maintain missing cases properly. The numeric scores on "rincom06"

and the appropriate midpoints are displayed in the image above. Note that for category 25, a true midpoint could not be constructed.

In the example above, the midpoint of each interval is used as an estimate; there are, however, other methods for estimating such intervals. Once you click "Continue" and then "OK" in the prior window, SPSS Statistics will create your newly coded variable, "rincomMP": respondent's income, range midpoint estimate.

Now, to carry out the one-way ANOVA test, use the following menus:

Analyze → Compare Means → One-Way ANOVA . . .

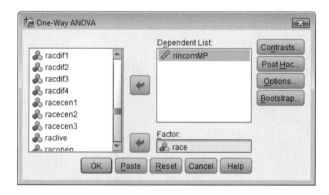

In the dialog box that is presented, choose the dependent and independent (factor) variables for your analysis. Click the "Options" key to customize your selections further; this will bring up the following dialog box:

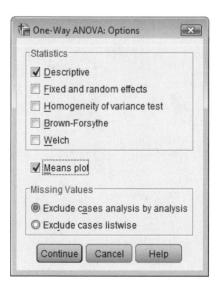

Here, descriptive statistics have been requested as well as a plot of the means. Cases have been excluded analysis by analysis. After clicking "Continue" here, and then "OK" in the "One-Way ANOVA" window, SPSS Statistics will run the test. The output that follows begins with the descriptive statistic information and is followed by the ANOVA results.

Descriptives

Range Midpoint Income Estimate

	N	Mean	Std. Deviation	Std. Error	95% Confidence Interval for Mean		Minimum	Maximum
					Lower Bound	Upper Bound		
WHITE	842	43128.8599	36391.52465	1254.13476	40667.2583	45590.4615	500.00	160000.00
BLACK	165	29071.2121	23930.82329	1863.01207	25392.6304	32749.7939	500.00	140000.00
OTHER	49	44612.2449	42969.31625	6138.47375	32270.0162	56954.4736	500.00	160000.00
Hispanic	146	24636.9863	25042.44814	2072.52775	20540.7190	28733.2536	500.00	140000.00
Total	1202	39013.5191	34741.02287	1002.05225	37047.5516	40979.4867	500.00	160000.00

ANOVA

Range Midpoint Income Estimate

	Sum of Squares	df	Mean Square	F	Sig.
Between Groups	62282205423.987	3	20760735141.329	17.929	.000
Within Groups	1387251137390.074	1198	1157972568.773		
Total	1449533342814.061	1201			

The F statistic has been calculated at 17.929 with 3 degrees of freedom. In this example, F is significant at the $p = 0.000$ level. The means plot has also been provided by SPSS Statistics as requested and has been reproduced below. Although the line itself may not be of much use here, as perhaps it might be in the case of an ordinal variable, it is easy to quickly ascertain the differences in the means from this graph.

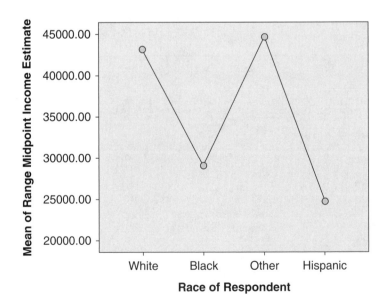

ANOVA in Regression

To examine ANOVA in an ordinary least squares (OLS) regression model, use the following menus:

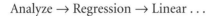

Analyze → Regression → Linear . . .

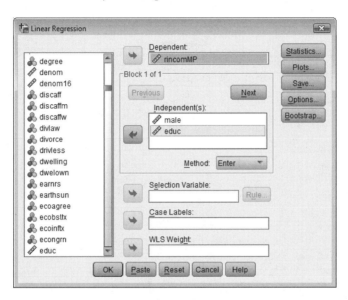

When presented with the "Linear Regression" dialog box, enter your dependent variable and independent variables. To obtain additional partial correlation values or estimates, click the "Statistics" button and mark the appropriate boxes in that dialog box.

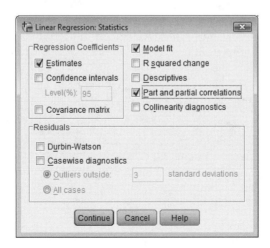

After you click "Continue" in the above dialog box and then "OK" in the "Linear Regression" dialog box, the following output will be presented. Note the sum of squares and mean squares have been provided, as well as the F statistic and its p value (indicating whether it is statistically significant).

Remember that r-square is a PRE (proportional reduction in error) statistic, and when $r^2 = 0.216$, we can interpret that to mean that 21.6% of the variation in income is explained by variations in education and gender.

Model Summary

Model	R	R Square	Adjusted R Square	Std. Error of the Estimate
1	.465[a]	.216	.215	30800.701

a. Predictors: (Constant), HIGHEST YEAR OF SCHOOL COMPLETED, Male

Below, you can see that F was calculated at 165.042 with 2 degrees of freedom. This value is statistically significant with $p = 0.000$.

ANOVA[a]

Model		Sum of Squares	df	Mean Square	F	Sig.
1	Regression	313145218575.747	2	156572609287.874	165.042	.000[b]
	Residual	1135573747622.159	1197	948683164.262		
	Total	1448718966197.906	1199			

a. Dependent Variable: Range Midpoint Income Estimate

b. Predictors: (Constant), HIGHEST YEAR OF SCHOOL COMPLETED, Male

As part of the linear regression command, SPSS Statistics also provides the regression model coefficient output, printed below. Notice that, on average, men are predicted to make almost $16,000 per year more than women, and each additional year of education is estimated to bring an average of just under $5,000 per year in annual earnings, not controlling for any other factors.

Coefficients[a]

Model		Unstandardized Coefficients		Standardized Coefficients			Correlations		
		B	Std. Error	Beta	t	Sig.	Zero-order	Partial	Part
1	(Constant)	-37478.523	4446.991		-8.428	.000			
	Male	15971.913	1785.136	.229	8.947	.000	.204	.250	.229
	HIGHEST YEAR OF SCHOOL ...	4939.869	302.587	.419	16.325	.000	.405	.427	.418

a. Dependent Variable: Range Midpoint Income Estimate

11

Editing Output

The SPSS Statistics Output Editor allows a great degree of freedom for editing charts, tables, and other types of output. This information can all be exported to other computer programs for inclusion in documents created and developed in those programs. In particular, the information and graphics from SPSS Statistics can be suitably imported and handled within Microsoft Word, as well as other computer word processing programs.

Editing Basic Tables

The first table below was produced by default by SPSS Statistics for a "Compare Means" function request for respondents' occupational prestige scores and sex variables, using the following procedure:

Analyze → Compare Means → Means ...

Report

RS OCCUPATIONAL PRESTIGE SCORE (1980)

RESPONDENTS SEX	Mean	N	Std. Deviation
MALE	43.74	836	13.547
FEMALE	43.36	1054	13.943
Total	43.53	1890	13.767

The next table has been edited. Labels have been added or edited, and at least one column was resized. You can directly edit the table in the SPSS Output Viewer window by double-clicking on the table and moving cells, double-clicking and retyping labels, and so on. This is done much the same way that table editing is done using a spreadsheet such as Microsoft Excel. Alternatively, you can use the following menu command:

Edit → Edit Content → In Viewer . . .

Prestige of Job by Gender (GSS, 2010)

RS OCCUPATIONAL PRESTIGE SCORE (1980)

RESPONDENTS SEX	Mean	N	Std. Deviation
MEN	43.74	836	13.547
WOMEN	43.36	1054	13.943
Total	43.53	1890	13.767

The table above was edited using the Table Properties option, located by right-clicking a cell in the table (or selecting <Control> + clicking the mouse for standard one-button mice on Apple Macintosh computers), as shown below:

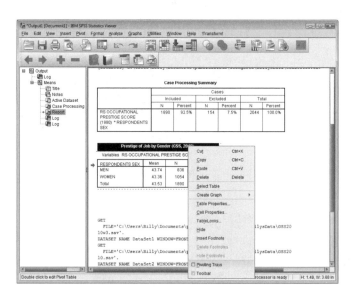

After making the Table Properties selection, you will be provided with the dialog box that follows, which will allow editing of borders, footnotes, formatting, and so on.

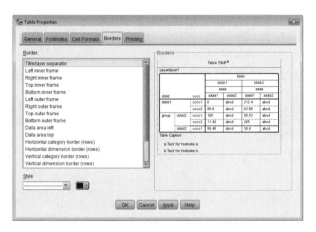

If you wish to have the table open in a new window for editing, keeping it separate from all of the other output in your viewer, you can use the following menu options. This can allow more focus on the table being edited.

Edit → Edit Content → In Separate Window . . .

You can also use the SPSS Statistics pivoting trays function to change the layout of the table if you wish. While the table has been selected for editing, as described above, use the following menus to reveal pivoting trays.

Pivot → Pivoting Trays

You still have direct access to changing the table in the Output Viewer or separate window, depending on which you have selected. The only difference is that in addition to editing the table itself, you now can avail yourself of the functions in the "Pivoting Trays" window.

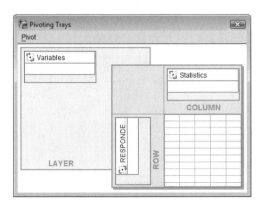

The SPSS Statistics "Pivoting Trays" window gives the user more direct control over the features of the table, including the structure of the variables. This is particularly useful when there are multiple variables in the analysis, including one or more control variables. Interchanging variables using pivoting trays is easy to do and easy to reverse.

Copying to Microsoft Word

There are several options for getting an SPSS Statistics table into Microsoft Word. One way is to select the table in the SPSS Output Viewer by clicking on it once. Then copy the table by choosing this menu:

Edit → Copy

The keyboard shortcut for that function is <Ctrl> + C on a Microsoft Windows PC and <Apple> + C on an Apple Macintosh computer.

Edit → Copy Special . . .

At this point, you can paste the output into your word processor (e.g., Microsoft Word). This can be done by selecting the following menus in Microsoft Word:

Edit → Paste

Newer versions of Microsoft Word (Office 2007 for Windows and Office 2008 for Macintosh) have a different menu structure. Simply click the "Paste" icon in the menu bar, or the word "Paste" underneath it. Again, there is a keyboard shortcut for this function: <Ctrl> + V on a Windows PC or <Apple> + V on an Apple computer.

By copying and pasting in this way, you will still be able to edit the tables in Microsoft Word. It can, however, be a bit more cumbersome to do the table editing in Word, and it can also pose layout complications. For instance, resizing the table may necessitate recalibrating font sizes within the table as well as many row and column dimensions. It usually benefits the user to do the table editing in SPSS Statistics and then, when editing is complete, copy the table into a document where it can be resized proportionally, like a photo or other object. To do this, select these menus in SPSS Statistics (and then those that follow in Microsoft Word or similar menus in another word processing program):

Copy → Special . . .

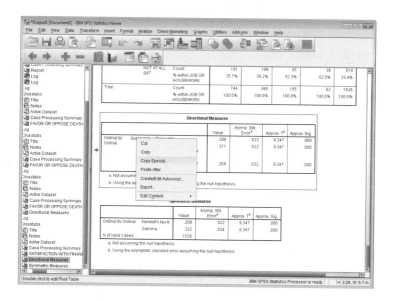

Or, you can right-click the table, as shown above and select "Copy Special . . ." that way. You will be presented with the following dialog box:

Make certain to check the box next to "Image." By initial default, this box is *not* selected. You can change the default by selecting the "Save as default" checkbox at the bottom of this window. Next, select these menu options in Microsoft Word (or similar menus in another word processing program):

Edit → Paste Special . . .

Then double-click "Picture" in the "Paste Special" box that appears. There may be several picture formats to choose from (e.g., JPG, PNG). Usually, any of these formats will work; JPG files tend to have the greatest range of compatibility with other software. This will paste the table into the document as a largely noneditable (but resizable) object. Although the internal characteristics of the table can no longer be changed at this point, placement and sizing of the table is much more convenient this way.

Newer versions of Microsoft Word (Office 2007 for Windows and Office 2008 for Macintosh) have a different menu structure. With the "Home" menu tab selected in Office 2007 for Windows, click the arrow under "Paste" in the upper left corner of the screen and choose "Paste Special" from the pull-down menu that appears.

Importing and Preparing Text Files for Analysis by SPSS

It is possible to have SPSS Statistics automatically import data from a text file that was created or edited with some other program that does not have a translation option when opening files, such as Microsoft Excel or Stata. For those files, you would simply open the file and SPSS would categorize and format the data automatically. With text files, there is a six-step process to do this.

File → Open Data . . .

When presented with the "Open Data" dialog box, click the pull-down menu for "Files of type:" and select "Text." These files may have name extensions such as *.txt, *.dat, or *.csv. If you have a text file that does not have one of these extensions, you might append the ".txt" to your file. Now, click "Open" and the Text Import Wizard will be launched.

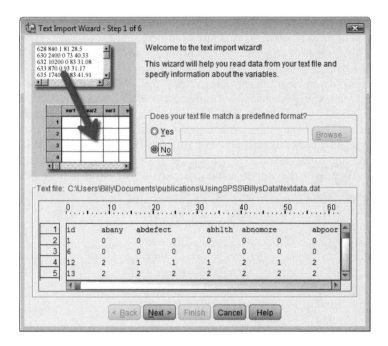

If your file does not have special formatting, click "No" where asked if your file matches a predefined format. Then, click "Next."

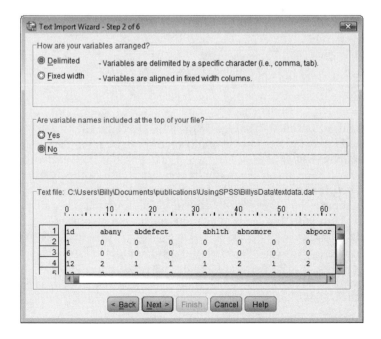

When presented with Step 2 of the Text Import Wizard, choose whether your data are separated into columns of exact width, or if there is something specific separating each number (such as a comma, a semi-colon, a space, or a tab). Also, if the first row of text in your data file includes the names of the variables, select "Yes." Once you do that, you will see the names of the variables disappear from the preview at the bottom of the window, indicating that the names have already been classified. Now, click "Next."

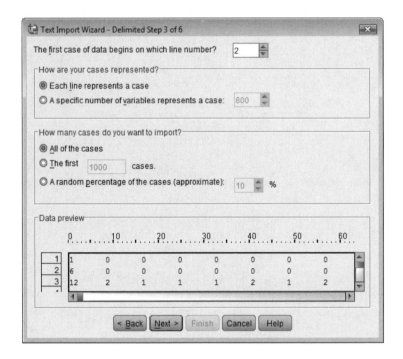

In this example, the data begin on Line 2 because the variable names occupied Line 1. If you have an additional space between the names and the data, you would choose Line 3 in the dialog box above. If your data are line by line, choose "Each line represents a case"; otherwise, if there are no returns in your data file (or if there are frequent returns within cases), choose the other option and indicate how many variables there are (total) in your file. Note that there must be a response (even if for missing data) for each variable in each case for this method to work. Finally, choose how many cases you wish to import. Typically, all cases are imported and then you can sample from them quickly and easily using SPSS Statistics (see Chapter 3: Selecting and Sampling Cases). Now, click "Next."

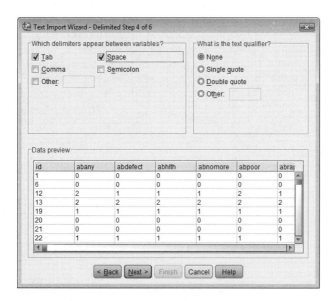

Choose the characters or delimiters that separate variables in your text file. Then determine whether text strings in your file are marked with quotes or other indicators. Now, click "Next."

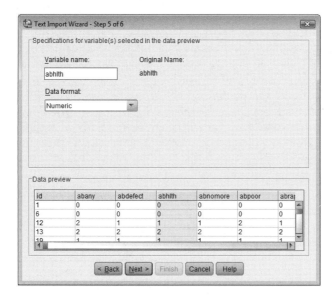

In Step 5 of the Text Import Wizard, you can rename any of your variables and establish the type (numeric, string, etc.). You could skip this step and carry out these housekeeping tasks in the Variable View of SPSS Statistics once the file has been successfully imported. Now, click "Next."

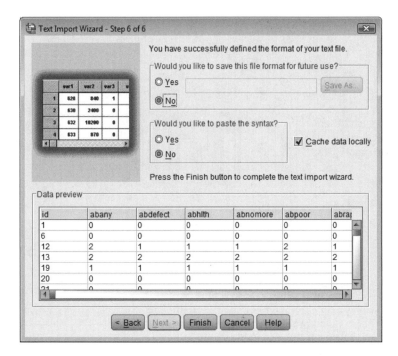

In the final step of the text importer, you are offered the option to save the setting for this import, which can be useful if you will be importing text files of this precise format in the future. Also, SPSS Statistics gives you the option to paste the command syntax for this function into the Output Viewer, which could also be used later to perform this task identically, or with edits for different variable names, and so on. Click "Finish" and you will be presented with the imported file in the Data Editor window.

Editing Charts and Graphs

The SPSS Statistics Output Viewer allows for interactive editing of charts and graphs. Not only can labels, titles, numbers, legends, and so forth be added and edited, but the actual type and style of the chart or graph can be changed from the output interface.

Suppose you create a clustered bar graph, like the one that is created below:

Graphs → Chart Builder . . .

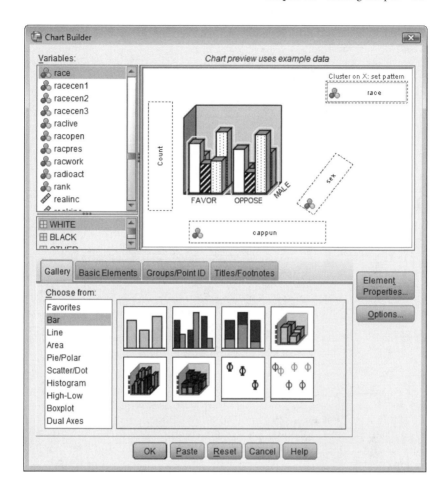

For this graph, we wish to examine the dependent variable "cappun" (opinion about capital punishment/favoring or opposing the death penalty for murder) as distributed across race/ethnicity and gender. We will use race and sex, respectively, to analyze those dimensions. This version of "race" is the recoded version, including information from the variable "Hispanic." For details on how this variable was recoded, see Chapter 2: Transforming Variables. The race recode is specifically dealt with in the following section of the chapter: "Recoding Using Two or More Variables to Create a New Variable."

Click "OK" to produce the bar graph. Now, for a more comprehensive and interactive way to edit the bar graph (or any other SPSS Statistics chart or graph), select and double-click the chart or graph. A new "Chart Editor" window will pop up, as pictured below:

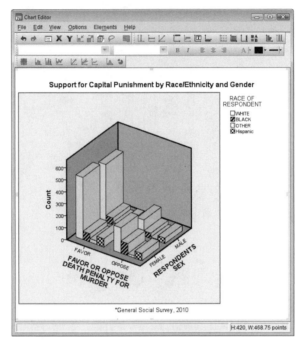

Note that the chart is shown above using patterns to differentiate columns, rather than color. Patterns are particularly useful for printing monochrome output, such as that produced on a black-and-white laser printer. Color charts are the default, unless you change the preferences settings for SPSS Statistics using the "Options" box, as seen below:

Edit → Options . . .

However, even if you have not changed the default from color to patterns, there is no need to scrap your work. You can switch from color to patterns (or the other way around) by using the Properties window. See the illustration of the Properties window after the next paragraph for details on exactly how to perform the switch.

Numerous editing options are available in the Properties window, which usually opens automatically with the "Chart Editor" but can also be accessed by selecting the following menus:

Edit → Properties

You can also reach the "Properties" box by using the shortcuts <Ctrl> + T on a Windows PC or <Apple> + T on a Macintosh computer.

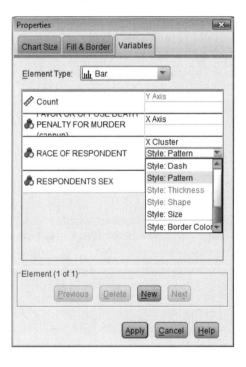

Using the pull-down menu for "Element Type," you can change the type of graph in the output, from a bar graph, for example, to some other type of graph, like a pie chart or an area graph. It is not necessary to go through and rerun the graph/chart function. It can be rebuilt directly from the "Chart Editor."

By selecting "Style" by the X Cluster variable, "race of respondent," you can choose whether to have the graph in color, black-and-white patterns, or some other method of illustration. For this example, the "pattern" style was chosen over the color option, as is often best for a monochrome printer output.

A title has been added by selecting Options → Title . . . from the "Chart Editor" window. Here, a text box will be added to the top area of the "Chart Editor" with the word "title"; simply replace that word with what you would like to title your chart. A footnote to indicate the source of data used to create the graph has been added by selecting Options → Footnote . . . Here, a text box will be added to the bottom area of the "Chart Editor" with the word "footnote"; replace this with the appropriate footnote. In this case, the data come from the 2010 General Social Survey.

Also, notice that the graph has been rotated vertically, so that we are able to better see the bars in the back row. You can change the vertical and horizontal rotations on the overall image by using the following menus:

Edit → 3-D Rotation

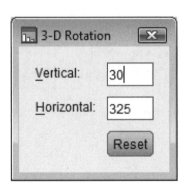

For this example, the vertical rotation was augmented to 30 degrees, but you should select whatever degree of rotation works best with your image.

12

Advanced Applications

Merging Data From Multiple Files

There are typically two ways a user wants to merge files with SPSS Statistics. One way is to combine two data files that contain the same variables, but consist of different cases (e.g., two or more waves of surveys completed by different people, but including the same information). Another way is to have additional variables to add to existing cases (e.g., second round of responses from the same respondents). Note that you must be doing this from the Data Editor window containing the file (active data set) to which you wish to add cases or variables.

First, suppose you want to add cases:

Data → Merge Files → Add Cases . . .

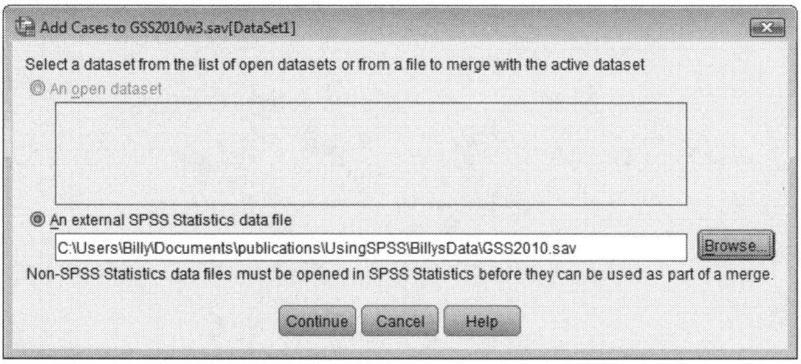

If you have already opened the data file in another SPSS Statistics Data Editor window, then you can select "An open dataset" and choose from the list that will appear. Otherwise, select "An external SPSS Statistics data file,"

and locate the file you wish to add. Both files, of course, need to be in SPSS (*.sav) format. If the file you are adding is not in SPSS data format, then you should first import that file into SPSS Statistics before carrying out the merge function. For more information about importing files, see Chapter 1 of this book.

You will be presented with a new dialog box, like the one that follows:

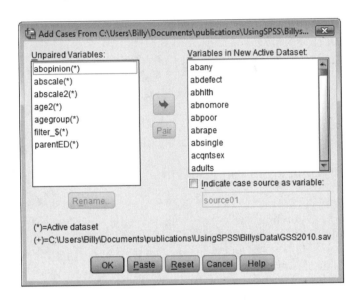

Assuming there are no unpaired variables, or you are not concerned with pairing variables (same variables in two different data sets each with a different name), select "OK," and SPSS Statistics will perform the addition of cases to your data file.

Now, suppose you want to add new variables to your data set:

File → Data → Add Variables ...

Again, select the appropriate file, whether it is open in another instance (window) of SPSS Statistics, or it is located on a disk or server connected to your computer. Click "Continue," and you will be presented with the next dialog box:

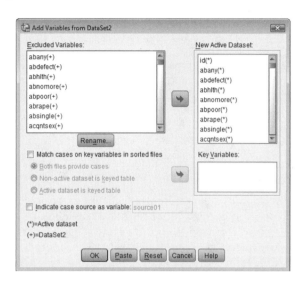

The "New Active Dataset" box shows the variables that will be contained in the newly merged data file. The excluded variables are those that are duplicates in name. If any of those are not duplicates, but happen to have the same name, then select each one and click the "Rename" button to correct the problem. If you are not certain that the cases are listed in the same order in the two data files, then you must match the cases by some identifying variable (e.g., a case ID): Select the variable that contains that information and move it into the "Key Variables" box. Click "OK," and SPSS Statistics will perform the merge as instructed.

Opening Previously Created Syntax Files

Syntax files provide computer code to instruct SPSS Statistics to perform functions. Most of these functions can be achieved by using the point-and-click method that this book uses. That is, functions can be performed by using the menus at the top of the Data Editor window. Creating syntax code using SPSS Statistics syntax computer language is not addressed in this book. If you have an SPSS Statistics syntax file, however, with code for performing SPSS commands, open it as follows:

File → Open → Syntax

You will get a dialog box like the one below:

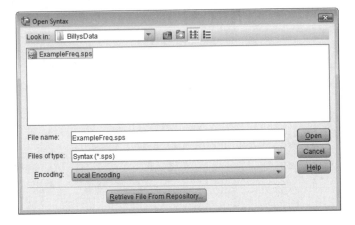

Navigate your computer files, locate your file, and open it. The syntax code will appear in a "Syntax Editor" window, like the one that follows. Select the portion of the code that you wish to run. (The example below runs frequency information on a data file that is supposed to already be opened.) Click the "run" button to execute the computer code. (The "run" button is the blue triangle.) You can also type <Ctrl> + R (<Apple> + R for Macintosh computers) or select from the menu:

Run → Selection

From the menu, you also have the option to run the whole file or to run it from a particular point to the end. The menus also afford the option to choose which lines (from, to) of syntax code to run.

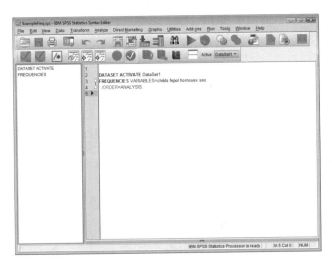

Creating New SPSS Syntax Files

Although this book will not provide detailed information about creating syntax files, there are a few things about syntax files that may be useful, even for those users who have no intention of writing code to perform SPSS Statistics functions. To create a new syntax file, select the following menu options:

File → New → Syntax

A new window, Syntax Editor, will appear. Information that is typed into this interface will form the syntax file. All or part of what is typed into this interface can be used either immediately or at a later time.

With any of the SPSS Statistics point-and-click functions—those operations implemented using the menus at the top of the editor windows—there is an option to select "Paste" instead of clicking "OK." When you do this, SPSS Statistics does not execute the function; instead, it records the instructions for performing the function in a syntax window. A user may choose to save the instructions for later use, or run them immediately and save a copy for future use or reference. What is pasted are the instructions that SPSS Statistics gives "behind the scenes" for that particular function (e.g., frequency distributions). Saving the commands in this way allows faster replication of a series of tasks.

Saving pasted SPSS Statistics syntax files can be useful for those who are performing many operations that are repetitive or similar across variables— particularly if they are more complicated functions. Take recoding variables, for instance. If you are recoding a number of variables with similar, but not identical, schemes or names, then by pasting and editing the same recode command for multiple variables, you could potentially save a great deal of time.

Saving syntax files also provides a complete record of how a data file was altered, which can be helpful to some users because SPSS Statistics will show the altered data file only and not provide a list of updates that have been made.

About the Author

William E. Wagner, III, PhD, is a Professor of Sociology at California State University, Channel Islands. Prior to coming to CSU, Channel Islands, he served as a member of the faculty and Director of the Institute for Social and Community Research at California State University, Bakersfield. He completed his PhD in sociology at the University of Illinois, Chicago. Dr. Wagner also holds an undergraduate degree in mathematics from St. Mary's College of Maryland. He has published in national and regional scholarly journals on topics such as urban sociology, homophobia, academic status, sports, and public health. Dr. Wagner is coauthor of the seventh edition of *Adventures in Social Research,* published by Pine Forge Press (2010) as well as the forthcoming eighth edition (2012) of *Adventures in Social Research* with Earl Babbie et al.

Dr. Wagner's webpage is http://faculty.csuci.edu/william.wagner.

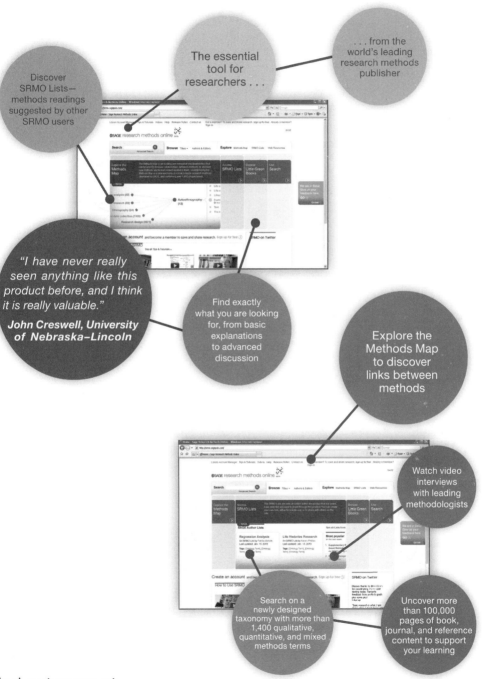

$SAGE research**methods**
The Essential Online Tool for Researchers

Discover SRMO Lists—methods readings suggested by other SRMO users

The essential tool for researchers . . .

. . . from the world's leading research methods publisher

"I have never really seen anything like this product before, and I think it is really valuable."

John Creswell, University of Nebraska–Lincoln

Find exactly what you are looking for, from basic explanations to advanced discussion

Explore the Methods Map to discover links between methods

Watch video interviews with leading methodologists

Search on a newly designed taxonomy with more than 1,400 qualitative, quantitative, and mixed methods terms

Uncover more than 100,000 pages of book, journal, and reference content to support your learning

find out more at
srmo.sagepub.com